The Fairy-Land of Science
Arabella B. Buckley

TABLE OF CONTENTS

Lecture I The Fairy-Land of Science; How to Enter It; How to Use It; And How to Enjoy It
Lecture II Sunbeams, and the Work They Do
Lecture III The Aerial Ocean in Which We Live
Lecture IV A Drop of Water on its Travels
Lecture V The Two Great Sculptors - Water and Ice
Lecture VI The Voices of Nature, and How We Hear Them
Lecture VII The Life of a Primrose
Lecture VIII The History of a Piece of Coal
Lecture IX Bees in the Hive
Lecture X Bees and Flowers

Week 1

LECTURE I

HOW TO ENTER IT; HOW TO USE IT; AND HOW TO ENJOY IT

I HAVE promised to introduce you today to the fairy-land of science - a somewhat bold promise, seeing that most of you probably look upon science as a bundle of dry facts, while fairy- land is all that is beautiful, and full of poetry and imagination. But I thoroughly believe myself, and hope to prove to you, that science is full of beautiful pictures, of real poetry, and of wonder-working fairies; and what is more, I promise you they shall be true fairies, whom you will love just as much when you are old and greyheaded as when you are young; for you will be able to call them up wherever you wander by land or by sea, through meadow or through wood, through water or through air; and though they themselves will always remain invisible, yet you will see their wonderful poet at work everywhere around you.

Let us first see for a moment what kind of tales science has to tell, and how far they are equal to the old fairy tales we all know so well. Who does not remember the tale of the "Sleeping Beauty in the Wood," and how under the spell of the angry fairy the maiden pricked herself with the spindle and slept a hundred years? How the horses in the stall, the dogs in the court-yard, the doves on the roof, the cook who was boxing the scullery boy's ears in the kitchen, and the king and queen with all their courtiers in the hall remained spell-bound, while a thick hedge grew up all round the castle and all within was still as death. But when the hundred years had passed the valiant prince came, the thorny hedge opened before him bearing beautiful flowers; and he, entering the castle, reached the room where the princess lay, and with one sweet kiss raised her and all around her to life again.

Can science bring any tale to match this?

Tell me, is there anything in this world more busy and active than water, as it rushes along in the swift brook, or dashes over the stones, or spouts up in the fountain, or trickles down from the roof, or shakes itself into ripples on the surface of the pond as the wind blows over it? But have you never seen this water spell-bound and motionless? Look out of the window some cold frosty morning in winter, at the little brook which yesterday was flowing gently past the house, and see how still it lies, with the stones over which it was dashing now held tightly in its icy grasp. Notice the wind-ripples on the pond; they have become fixed and motionless. Look up at the roof of the house. There, instead of living doves merely charmed to sleep, we have running water caught in the very act of falling and turned into transparent icicles, decorating the eaves with a beautiful crystal fringe. On every tree and bush you will catch the water- drops napping, in the form of tiny crystals; while the fountain looks like a tree of glass with long down-hanging pointed leaves. Even the damp of your own breath lies rigid and still on the window-pane frozen into delicate patterns like fern-leaves of ice.

All this water was yesterday flowing busily, or falling drop by drop, or floating invisibly in the air; now it is all caught and spell-bound - by whom? By the enchantments of the frost-giant who holds it fast in his grip and will not let it go.

But wait awhile, the deliverer is coming. In a few weeks or days, or it may be in a few hours, the brave sun will shine down; the dull-grey, leaden sky will melt before his, as the hedge gave way before the prince in the fairy tale, and when the sunbeam gently kisses the frozen water it will be set free. Then the brook will flow rippling on again; the frost-drops will be shaken down from the trees, the icicles fall from the roof, the moisture trickle down the window-pane, and in the bright, warm sunshine all will be alive again.

Is not this a fairy tale of nature? and such as these it is which science tells.

Again, who has not heard of Catskin, who came out of a hollow tree, bringing a walnut containing three beautiful dresses - the first glowing as the sun, the second pale and beautiful as the moon, the third spangled like the star-lit sky, and each so fine and delicate that all three could be packed into a walnut shell; and each one of these tiny structures is not the mere dress but the home of a living animal. It is a tiny, tiny shell-palace made of the most delicate lacework, each pattern being more beautiful than the last; and what is more, the minute creature that lives in it has built it out of the foam of the sea, though he himself is nothing more than a drop of jelly.

Lastly, anyone who has read the 'Wonderful Travellers' must recollect the man whose sight was so keen that he could hit the eye of a fly sitting on a tree two miles away. But tell me, can you see gas before it is lighted, even when it is coming out of the gas-jet close to your eyes? Yet, if you learn to use that wonderful instrument the spectroscope, it will enable you to tell one kind of gas from another, even when they are both ninety-one millions of miles away on the face of the sun; nay more, it will read for you the nature of the different gases in the far distant stars, billions of miles away, and actually tell you whether you could find there any of the same metals which we have on the earth.

We might find hundreds of such fairy tales in the domain of science, but these three will serve as examples, and we much pass on to make the acquaintance of the science-fairies themselves, and see if they are as real as our old friends.

Tell me, why do you love fairy-land? what is its charm? Is it not that things happen so suddenly, so mysteriously, and without man having anything to do with it? In fairy-land, flowers blow, houses spring up like Aladdin's palace in a single night, and people are carried hundreds of miles in an instant by the touch of a fairy wand.

And then this land is not some distant country to which we can never hope to travel. It is here in the midst of us, only our eyes must be opened or we cannot see it. Ariel and Puck did not live in some unknown region. On the contrary, Ariel's song is

"Where the bee sucks, there suck I;
In a cowslip's bell I lie;
There I couch when owls do cry.
On the bat's back I do fly,
After summer, merrily."

The peasant falls asleep some evening in a wood and his eyes are opened by a fairy wand, so that he sees the little goblins and imps dancing around him on the green sward, sitting on mushrooms, or in the heads of the flowers, drinking out of acorn-cups, fighting with blades of grass, and riding on grasshoppers.

So, too, the gallant knight, riding to save some poor oppressed maiden, dashes across the foaming torrent; and just in the middle, as he is being swept away, his eyes are opened, and he sees fairy water-nymphs soothing his terrified horse and guiding him gently to the opposite shore.

They are close at hand, these sprites, to the simple peasant or the gallant knight, or to anyone who has the gift of the fairies and can see them. but the man who scoffs at them, and does not believe in them or care for them, he never sees them. Only now and then they play him an ugly trick, leading him into some treacherous bog and leaving him to get out as he may.

Now, exactly all this which is true of the fairies of our childhood is true too of the fairies of science. There are forces around us, and among us, which I shall ask you to allow me to call fairies, and these are ten thousand times more wonderful, more magical, and more beautiful in their work, than those of the old fairy tales. They, too, are invisible, and many people live and die without ever seeing them or caring to see them. These people go about with their eyes shut, either because they will not open them, or because no one has taught them how to see. They fret and worry over their own little work and their own petty troubles, and do not know how to rest and refresh themselves, by letting the fairies open their eyes and show them the calm sweet picture of nature. They are like Peter Bell of whom Wordsworth wrote:-

"A primrose by a river's brim
A yellow primrose was to him,
And it was nothing more."

But we will not be like these, we will open our eyes and ask,
"What are these forces or fairies, and how can we see them?"
Just go out into the country, and sit down quietly and watch nature at work. Listen to the wind as it blows, look at the clouds rolling overhead, and waves rippling on the pond at your feet. Hearken to the brook as it flows by, watch the flower-buds opening one by one, and then ask yourself, "How all this is done?" Go out in the evening and see the dew gather drop by drop upon the grass, or trace the delicate hoar-frost crystals which bespangle every blade on a winter's morning. Look at the vivid flashes of lightening in a storm, and listen to the pealing thunder: and then tell me, by what machinery is all this wonderful work done? Man does none of it, neither could he stop it if he were to try; for it is all the work of those invisible forces or fairies whose acquaintance I wish you to make. Day and night, summer and winter, storm or calm, these fairies are at work, and we may hear them and know them, and make friends of them if we will.

There is only one gift we must have before we can learn to know them - we must have imagination. I do not mean mere fancy, which creates unreal images and impossible monsters, but imagination, the power of making pictures or images in our mind, of that which is, though it is invisible to us. Most children have this glorious gift, and love to picture to themselves all that is told them, and to hear the same tale over and over again till they see every bit of it as if it were real. This is why they are sure to love science it its tales are told them aright; and I, for one, hope the day may never come when we may lose that childish clearness of vision, which enables us through the temporal things which are seen, to realize those eternal truths which are unseen.

If you have this gift of imagination come with me, and in these lectures we will look for the invisible fairies of nature.

Watch a shower of rain. Where do the drops come from? and why are they round, or rather slightly oval? In our fourth lecture we shall se that the little particles of water of which the raindrops are made, were held apart and invisible in the air by heat, one of the most wonderful of our forces* or fairies, till the cold wind passed by and chilled the air. Then, when there was no longer so much heat, another invisible force, cohesion, which is always ready and waiting, seized on the tiny particles at once, and locked them together in a drop, the closest form in which they could lie. Then as the drops became larger and larger they fell into the grasp of another invisible force, gravitation, which dragged them down to the earth, drop by drop, till they made a shower of rain. Pause for a moment and think. You have surely heard of gravitation, by which the sun holds the earth and the planets, and keeps them moving round him in regular order? Well, it is this same gravitation which is a t work also whenever a shower of rain falls to the earth. Who can say that he is not a great invisible giant, always silently and invisibly toiling in great things and small whether we wake or sleep?
*(I am quite aware of the danger incurred by using this word "force", especially in the plural; and how even the most modest little book may suffer at the hands of scientific purists by employing it rashly. As, however, the better term "energy" would not serve here, I hope I may be forgiven for retaining the much- abused term, especially as I sin in very good company.)
Now the shower is over, the sun comes out and the ground is soon as dry as though no rain had fallen. Tell me; what has become of the rain-drops? Part no doubt have sunk into the ground, and as for the rest, why you will say the sun has dried them up. Yes, but how? The sun is more than ninety-one millions of miles away; how has he touched the rain-drops? Have you ever heard that invisible waves are travelling every second over the space between the sun and us? We shall see in the next lecture how these waves are the sun's messengers to the earth, and how they tear asunder the rain-drops on the ground, scattering them in tiny particles too small for us to see, and bearing them away to the clouds. Here are more invisible fairies working every moment around you, and you cannot even look out of the window without seeing the work they are doing.
If, however, the day is cold and frosty, the water does not fall in a shower of rain; it comes down in the shape of noiseless snow. Go out after such a snow-shower, on a calm day, and look at some of the flakes which have fallen; you will see, if you choose good specimens, that they are not mere masses of frozen water, but that each one is a beautiful six-pointed crystal star. How have these crystals been built up? What power

has been at work arranging their delicate forms? In the fourth lecture we shall see that up in the clouds another of our invisible fairies, which, for want of a better name, we call the "force of crystallization," has caught hold of the tiny particles of water before "cohesion" had made them into round drops, and there silently but rapidly, has moulded them into those delicate crystal starts know as "snowflakes".

And now, suppose that this snow-shower has fallen early in February; turn aside for a moment from examining the flakes, and clear the newly-fallen snow from off the flower-bed on the lawn. What is this little green tip peeping up out of the ground under the snowy covering? It is a young snowdrop plant. Can you tell me why it grows? where it finds its food? what makes it spread out its leaves and add to its stalk day by day? What fairies are at work here?

First there is the hidden fairy "life," and of her even our wisest men know but little. But they know something of her way of working, and in Lecture VII we shall learn how the invisible fairy sunbeams have been buy here also; how last year's snowdrop plant caught them and stored them up in it's bulb, and how now in the spring, as soon as warmth and moisture creep down into the earth, these little imprisoned sun-waves begin to be active, stirring up the matter in the bulb, and making it swell and burst upwards till it sends out a little shoot through the surface of the soil. Then the sun-waves above-ground take up the work, and form green granules in the tiny leaves, helping them to take food out of the air, while the little rootlets below are drinking water out of the ground. The invisible life and invisible sunbeams are busy here, setting actively to work another fairy, the force of "chemical attraction," and so the little snowdrop plant grows and blossoms, without any help from you or me.

Week 2

One picture more, and then I hope you will believe in my fairies. From the cold garden, you run into the house, and find the fire laid indeed in the grate, but the wood dead and the coals black, waiting to be lighted. You strike a match, and soon there is a blazing fire. Where does the heat come from? Why do the coals burn and give out a glowing light? Have you not read of gnomes buried down deep in the earth, in mines, and held fast there till some fairy wand has released them, and allowed them to come to earth again? Well, thousands and millions of years ago, those coals were plants; and like the snowdrop in the garden of to-day, they caught the sunbeams and worked them into their leaves. Then the plants died and were buried deep in the earth and the sunbeams with them; and like the gnomes they lay imprisoned till the coals were dug out by the miners, and brought to your grate; and just now you yourself took hold of the fairy wand which was to release them. You struck a match, and its atoms clashing with atoms of oxygen in the air, set the invisible fairies "heat" and "chemical attraction" to work, and they were soon busy within the wood and the coals causing their atoms too to clash; and the sunbeams, so long imprisoned, leapt into flame. Then you spread out your hands and cried, "Oh, how nice and warm!" and little thought that you were warming yourself with the sunbeams of ages and ages ago.

This is no fancy tale; it is literally true, as we shall see in Lecture VIII, that the warmth of a coal fire could not exist if the plants of long ago had not used the sunbeams to make their leaves, holding them ready to give up their warmth again whenever those crushed leaves are consumed.

Now, do you believe in, and care for, my fairy-land? Can you see in your imagination fairy 'Cohesion' ever ready to lock atoms together when they draw very near to each other: or fairy 'Gravitation' dragging rain-drops down to the earth: or the fairy of 'Crystallization' building up the snow-flakes in the clouds? Can you picture tiny sunbeam-waves of light and heat travelling from the sun to the earth? Do you care to know how another strange fairy, 'Electricity,' flings the lightning across the sky and causes the rumbling thunder? Would you like to learn how the sun makes pictures of the world on which he shines, so that we can carry about with us photographs or sun-pictures of all the beautiful scenery of the earth? And have you any curiosity about 'Chemical action,' which works such wonders in air, and land, and sea? If you have any wish to know and make friends of these invisible forces, the next question is

How are you to enter the fairy-land of science?

There is but one way. Like the knight or peasant in the fairy tales, you must open you eyes. There is no lack of objects, everything around you will tell some history if touched with the fairy wand of imagination. I have often thought, when seeing some sickly child drawn along the street, lying on its back while other children romp and play, how much happiness might be given to sick children at home or in hospitals, if only they were told the stories which lie hidden in the things around them. They need not even move from their beds, for sunbeams can fall on them there, and in a sunbeam there are stories enough to occupy a month. The fire in the grate, the lamp by the bedside, the water in the tumbler, the fly on the ceiling above, the flower in the vase on the table, anything, everything, has its history, and can reveal to us nature's invisible fairies.

Only you must with to see them. If you go through the world looking upon everything only as so much to eat, to drink, and to use, you will never see the fairies of science. But if you ask yourself why things happen, and how the great God above us has made and governs this world of ours; If you listen to the wind, and care to learn why it blows; if you ask the little flower why it opens in the sunshine and closes in the storm; and if when you find questions you cannot answer, you will take the trouble to hunt out in books, or make experiments to solve your own questions, then you will learn to know and love those fairies.

Mind, I do not advise you to be constantly asking questions of other people; for often a question quickly answered is quickly forgotten, but a difficulty really hunted down is a triumph for ever. For example, if you ask why the rain dries up from the ground, most likely you will be answered, "that the sun dries it," and you will rest satisfied with the sound of the words. But if you hold a wet handkerchief before the fire and see the damp rising out of it, then you have some real idea how moisture may be drawn up by heat from the earth.

A little foreign niece of mine, only four years old, who could scarcely speak English plainly, was standing one morning near the bedroom window and she noticed the damp trickling down the window-pane. "Auntie," she said, "what for it rain inside?" It was quite useless to explain to her in words, how our breath had condensed into drops of water upon the cold glass; but I wiped the pane clear, and breathed on it several times. When new drops were formed, I said, "Cissy and auntie have done like this all night in the room." She nodded her little head and amused herself for a long time breathing on the window-pane and watching the tiny drops; and about a month later, when we were travelling back to Italy, I saw her following the drops on the carriage window with her little finger, and heard her say quietly to herself, "Cissy and auntie made you." Had not even this little child some real picture in her mind of invisible water coming from her mouth, and making drops upon the window-pane?

Then again, you must learn something of the language of science. If you travel in a country with no knowledge of its language, you can learn very little about it: and in the same way if you are to go to books to find answers to your questions, you must know something of the language they speak. You need not learn hard scientific names, for the best books have the fewest of these, but you must really understand what is meant by ordinary words.

For example, how few people can really explain the difference between a solid, such as the wood of the table; a liquid, as water; and a gas, such as I can let off from this gas-jet by turning the tap. And yet any child can make a picture of this in his mind if only it has been properly put before him.

All matter in the world is made up of minute parts or particles; in a solid these particles are locked together so tightly that you must tear them forcibly apart if you with to alter the shape of the solid piece. If I break or bend this wood I have to force the particles to move round each other, and I have great difficulty doing it. But in a liquid, though the particles are still held together, they do not cling so tightly, but are able to roll or glide round each other, so that when you pour water out of a cup on to a table, it loses its cuplike shape and spreads itself out flat. Lastly, in a gas the particles are no longer held together at all, but they try to fly away from each other; and unless you shut a gas in tightly and safely, it will soon have spread all over the room.

A solid, therefore, will retain the same bulk and shape unless you forcibly alter it; a liquid will retain the same bulk, but no the same shape if it be left free; a gas will not retain either the same bulk or the same shape, but will spread over as large a space as it can find wherever it can penetrate. Such simple things as these you must learn from books and by experiment.

Then you must understand what is meant by chemical attraction; and though I can explain this roughly here, you will have to make many interesting experiments before you will really learn to know this wonderful fairy power. If I dissolve sugar in water, though it disappears it still remains sugar, and does not join itself to the water. I have only to let the cup stand till the water dries, and the sugar will remain at the bottom. There has been no chemical attraction here.

But now I will put something else in water which will call up the fairy power. Here is a little piece of the metal potassium, one of the simple substances of the earth; that is to say, we cannot split it up into other substances, wherever we find it, it is always the same. Now if I put this piece of potassium on the water it does not disappear quietly like the sugar. See how it rolls round and round, fizzing violently with a blue flame burning round it, and at last goes off with a pop.

What has been happening here?

You must first know that water is made of two substances, hydrogen and oxygen, and these are not merely held together, but are joined to completely that they have lost themselves and have become water; and each atom of water is made of two atoms of hydrogen and one of oxygen. Now the metal potassium is devotedly fond of oxygen, and the moment I threw it on the water it called the fairy "chemical attraction' to help it, and dragged the atoms of oxygen out of the water and joined them to itself. In doing this it also caught part of the hydrogen, but only half, and so the rest was left out in the cold. No, not in the cold! for the potassium and oxygen made such a great heat in clashing together that the rest of the hydrogen became very hot indeed, and sprang into the air to find some other companion to make up for what it had lost. Here it found some free oxygen floating about, and it seized upon it so violently, that they made a burning flame, while the potassium with its newly found oxygen and hydrogen sank down quietly into the water as potash. And so you see we have got quite a new substance potash in the basin; made with a great deal of fuss by chemical attraction drawing different atoms together.

When you can really picture this power to yourself it will help you very much to understand what you read and observe about nature.

Next, as plants grow around you on every side, and are of so much importance in the world, you must also learn something of the names of the different parts of a flower, so that you may understand those books which explain how a plant grows and lives and forms its seeds. You must also know the common names of the parts of an animal, and of your own body, so that you may be interested in understanding the use of the different organs; how you breathe, and how your blood flows; how one animal walks, another flies, and another swims. Then you must learn something of the various parts of the world, so that you may know what is meant by a river, a plain, a valley, or a delta. All these things are not difficult, you can learn them pleasantly from simple books on physics, chemistry, botany, physiology, and physical geography; and when you understand a few plain scientific terms, then all by yourself, if you will open your eyes and ears, you may wander happily in the fairy-land of science. Then wherever you go you will find

"Tongues in trees, books in the running brooks
Sermons in stones, and good in everything."

And now we come to the last part of our subject. When you have reached and entered the gates of science, how are you to use and enjoy this new and beautiful land?

This is a very important question for you may make a twofold use of it. If you are only ambitious to shine in the world, you may use it chiefly to get prizes, to be at the top of your class, or to pass in examinations; but if you also enjoy discovering its secrets, and desire to learn more and more of nature and to revel in dreams of its beauty, then you will study science for its own sake as well. Now it is a good thing to win prizes and be at the top of your class, for it shows that you are industrious; it is a good thing to pass well in examinations , for it show that you are accurate; but if you study science for this reason only, do not complain if you find it full, and dry, and hard to master. You may learn a great deal that is useful, and nature will answer you truthfully if you ask you questions accurately, but she will give you dry facts, just such as you ask for. If you do not love her for herself she will never take you to her heart.

This is the reason why so many complain that science is dry and uninteresting. They forget that though it is necessary to learn accurately, for so only we can arrive at truth, it is equally necessary to love knowledge and make it lovely to those who learn, and to do this we must get at the spirit which lies under the facts. What child which loves its mother's face is content to know only that she has brown eyes, a straight nose, a small mouth, and hair arranged in such and such a manner? No, it knows that its mother has the sweetest smile of any woman living; that her eyes are loving, her kiss is sweet, and that when she looks grave, then something is wrong which must be put right. And it is in this way that those who wish to enjoy the fairy-land of science must love nature.

It is well to know that when a piece of potassium is thrown on water the change which takes place is expressed by the formula K + H2O = KHO + H. But it is better still to have a mental picture of the tiny atoms clasping each other, and mingling so as to make a new substance, and to feel how wonderful are the many changing forms of nature. It is useful to be able to classify a flower and to know that the buttercup belongs to the Family Ranunculaceae, with petals free and definite, stamens hypogynous and indefinite, pistil apocarpous. But it is far sweeter to learn about the life of the little plant, to understand why its peculiar flower is useful to it, and how it feeds itself, and makes its seed. No one can love dry facts; we must clothe them with real meaning and love the truths they tell, if we wish to enjoy science.

Let us take an example to show this. I have here a branch of white coral, a beautiful, delicate piece of nature's work. We will begin by copying a description of it from one of those class-books which suppose children to learn words like parrots, and to repeat them with just as little understanding.

"Coral is formed by an animal belonging to the kingdom of Radiates, sub-kingdom Polypes. The soft body of the animal is attached to a support, the mouth opening upwards in a row of tentacles. The coral is secreted in the body of the polyp out of the carbonate of lime in the sea. Thus the coral animalcule rears its polypidom or rocky structure in warm latitudes, and constructs reefs or barriers round islands. It is limited in rage of depth from 25 to 30 fathoms. Chemically considered, coral is carbonate of like; physiologically, it is the skeleton of an animal; geographically, it is characteristic of warm latitudes, especially of the Pacific Ocean." This description is correct, and even fairly complete, if you know enough of the subject to understand it. But tell me, does it lead you to love my piece of coral? Have you any picture in your mind of the coral animal, its home, or its manner of working?

But now, instead of trying to master this dry, hard passage, take Mr.

Huxley's penny lecture on 'Coral and Coral Reefs,' and with the piece of coral in your hand try really to learn its history. You will then be able to picture to yourself the coral animal as a kind of sea-anemone, something like those which you have often seen, like red, blue, or green flowers, putting out feelers in sea-water on our coasts, and drawing in the tiny sea-animals to digest them in that bag of fluid which serves the sea-anemone as a stomach. You will learn how this curious jelly animal can split itself in two, and so form two polyps, or send a bud out of its side and so grow up into a kind of "tree or bush of polyps," or how it can hatch little eggs inside it and throw out young ones from its mouth, provided with little hairs, by means of which they swim to new resting-places. You will learn the difference between the animal which builds up the red coral as its skeleton, and the group of animals which build up the white; and you will look with new interest on our piece of white coral, as you read that each of those little sups on its stem with delicate divisions like the spokes of a wheel has been the home of a separate polyp, and that from the sea-water each little jelly animal has drunk in carbonate of lime as you drink in sugar dissolved in water, and then has used it grain by grain to build that delicate cup and add to the coral tree.

We cannot stop to examine all about coral now, we are only learning how to learn, but surely our specimen is already beginning to grow interesting; and when you have followed it out into the great Pacific Ocean, where the wild waves dash restlessly against the coral trees, and have seen these tiny drops of jelly conquering the sea and building huge walls of stone against the rough breakers, you will hardly rest till you know all their history. Look at that curious circular island in the picture, covered with palm trees; it has a large smooth lake in the middle, and the bottom of this lake is covered with blue, red, and green jelly animals, spreading out their feelers in the water and looking like beautiful flowers, and all round the outside of the island similar animals are to be seen washed by the sea waves. Such islands as this have been build entirely by the coral animals, and the history of the way in which the reefs have sunk gradually down, as the tiny creatures added to them inch by inch, is as fascinating as the story of the building of any fairy palace in the days of old. Read all this, and then if you have no coral of your own to examine, go to the British Museum and see the beautiful specimens in the glass cases there, and think that they have been built up under the rolling surf by the tiny jelly animals; and then coral will become a real living thing to you, and you will love the thoughts it awakens.

But people often ask, what is the use of learning all this? If you do not feel by this time how delightful it is to fill your mind with beautiful pictures of nature, perhaps it would be useless to say more. But in this age of ours, when restlessness and love of excitement pervade so many lives, is it nothing to be taken out of ourselves and made to look at the wonders of nature going on around us? Do you never feel tired and "out of sorts," and want to creep away from your companions, because they are merry and you are not? Then is the time to read about the starts, and how quietly they keep their course from age to age; or to visit some little flower, and ask what story it has to tell; or to watch the clouds, and try to imagine how the winds drive them across the sky. No person is so independent as he who can find interest in a bare rock, a drop of water, the foam of the sea, the spider on the wall, the flower underfoot or the starts overhead. And these interests are open to everyone who enters the fairy-land of science.

Moreover, we learn from this study to see that there is a law and purpose in everything in the Universe, and it makes us patient when we recognize the quiet noiseless working of nature all around us. Study light, and learn how all colour, beauty, and life depend on the sun's rays; note the winds and currents of the air, regular even in their apparent irregularity, as they carry heat and moisture all over the world. Watch the water flowing in deep quiet streams, or forming the vast ocean; and then reflect that every drop is guided by invisible forces working according to fixed laws. See plants springing up under the sunlight, learn the secrets of plant life, and how their scents and colours attract the insects. Read how insects cannot live without plants, nor plants without the flitting butterfly or the busy bee. Realize that all this is worked by fixed laws, and that out of it (even if sometimes in suffering and pain) springs the wonderful universe around us. And then say, can you fear for your own little life, even though it may have its troubles? Can you help feeling a part of this guided and governed nature? or doubt that the power which fixed the laws of the stars and of the tiniest drop of water - that made the plant draw power from the sun, the tine coral animal its food from the dashing waves; that adapted the flower to the insect and the insect to the flower - is also moulding your life as part of the great machinery of the universe, so that you have only to work, and to wait, and to love?

We are all groping dimly for the Unseen Power, but no one who loves nature and studies it can ever feel alone or unloved in the world. Facts, as mere facts, are dry and barren, but nature is full of life and love, and her calm unswerving rule is tending to some great though hidden purpose. You may call this Unseen Power what you will - may lean on it in loving, trusting faith, or bend in reverent and silent awe; but even the little child who lives with nature and gazes on her with open eye, must rise in some sense or other through nature to nature's God.

Week 3

Lecture II Sunbeams and How They Work

Who does not love the sunbeams, and feel brighter and merrier as he watches them playing on the wall, sparkling like diamonds on the ripples of the sea, or making bows of coloured light on the waterfall? Is not the sunbeam so dear to us that it has become a household word for all that is merry and gay? and when we want to describe the dearest, busiest little sprite amongst us, who wakes a smile on all faces wherever she goes, do we not call her the "sunbeam of the house"?

And yet how little even the wisest among us know about the nature and work of these bright messengers of the sun as they dart across space!

Did you ever wake quite early in the morning, when it was pitch- dark and you could see nothing, not even your own hand; and then lie watching as time went on till the light came gradually creeping in at the window? If you have done this you will have noticed that you can at first only just distinguish the dim outline of the furniture; then you can tell the difference between the white cloth on the table and the dark wardrobe beside it; then by degrees all the smaller details, the handles of the drawer, the pattern on the wall, and the different colours of all the objects in the room become clearer and clearer till at last you see all distinctly in broad daylight.

What has been happening here? and why have the things in the room become visible by such slow degrees? We say that the sun is rising, but we know very well that it is not the sun which moves, but that our earth has been turning slowly round, and bringing the little spot on which we live face to face with the great fiery ball, so that his beams can fall upon us.

Take a small globe, and stick a piece of black plaster over England, then let a lighted lamp represent the sun, and turn the globe slowly, so that the spot creeps round from the dark side away from the lamp, until it catches, first the rays which pass along the side of the globe, then the more direct rays, and at last stands fully in the blaze of the light. Just this was happening to our spot of the world as you lay in bed and saw the light appear; and we have to learn today what those beams are which fall upon us and what they do for us.

First we must learn something about the sun itself, since it is the starting-place of all the sunbeams. If the sun were a dark mass instead of a fiery one we should have none of these bright cheering messengers, and though we were turned face to face with him every day we should remain in one cold eternal night. Now you will remember we mentioned in the last lecture that it is heat which shakes apart the little atoms of water and makes them gloat up in the air to fall again as rain; and that if the day is cold they fall as snow, and all the water is turned into ice. But if the sun were altogether dark, think how bitterly cold it would be; far colder than the most wintry weather ever known, because in the bitterest night some warmth comes out of the earth, where it has been stored from the sunlight which fell during the day. But if we never received any warmth at all, no water would ever rise up into the sky, no rain ever fall, no rivers flow, and consequently no plants could grow and no animals live. All water would be in the form of snow and ice, and the earth would be one great frozen mass with nothing moving upon it.

So you see it becomes very interesting for us to learn what the sun is, and how he sends us his beams. How far away from us do you think he is? On a fine summer's day when we can see him clearly, it looks as if we had only to get into a balloon and reach him as he sits in the sky, and yet we know roughly that he is more than ninety-one millions of miles distant from our earth.

These figures are so enormous that you cannot really grasp them. But imagine yourself in an express train, travelling at the tremendous rate of sixty miles an hour and never stopping. At that rate, if you wished to arrive at the sun today you would have been obliged to start 171 years ago. That is, you must have set off in the early part of the reign of Queen Anne, and you must have gone on, never, never resting, through the reigns of George I, George ii, and the long reign of George III, then through those of George IV, William IV, and Victoria, whirling on day and night at express speed, and at last, today, you would have reached the sun!

And when you arrived there, how large do you think you would find him to be? Anaxagoras, a learned Greek, was laughed at by all his fellow Greeks because he said that the sun was as large as the Peloponne-sus, that is about the size of Middlesex. How astonished they would have been if they could have known that not only is he bigger than the whole of Greece, but more than a million times bigger than the whole world!

Our world itself is a very large place, so large that our own country looks only like a tiny speck upon it, and an express train would take nearly a month to travel round it. Yet even our whole globe is nothing in size compared to the sun, for it only measures 8000 miles across, while the sun measures more the 852,000.

Imagine for a moment that you could cut the sun and the earth each in half as you would cut an apple; then if you were to lay the flat side of the half-earth on the flat side of the half sun it would take 106 such earths to stretch across the face of the sun. One of these 106 round spots on the diagram represents the size which our earth would look if placed on the sun; and they are so tiny compared to him that they look only like a

string of minute beads stretched across his face. Only think, then, how many of these minute dots would be required to fill the whole of the inside of Fig. 4, if it were a globe.
One of the best ways to form an idea of the whole size of the sun is to imagine it to be hollow, like an air-ball, and then see how many earths it would take to fill it. You would hardly believe that it would take one million, three hundred and thirty-one thousand globes the size of our world squeezed together. Just think, if a huge giant could travel all over the universe and gather worlds, all as big as ours, and were to make first a heap of merely ten such worlds, how huge it would be! Then he must have a hundred such heaps of ten to make a thousand world; and then he must collect again a thousand times that thousand to make a million, and when he had stuffed them all into the sun-ball he would still have only filled three-quarters of it!
After hearing this you will not be astonished that such a monster should give out an enormous quantity of light and heat; so enormous that it is almost impossible to form any idea of it. Sir John Herschel has, indeed, tried to picture it for us. He found that a ball of lime with a flame of oxygen and hydrogen playing round it (such as we use in magic lanterns and call oxy- hydrogen light) becomes so violently hot that it gives the most brilliant artificial light we can get - such that you cannot put your eye near it without injury. Yet if you wanted to have a light as strong as that of our sun, it would not be enough to make such a lime-ball as big as the sun is. No, you must make it as big as 146 suns, or more than 146,000,000 times as big as our earth, in order to get the right amount of light. Then you would have a tolerably good artificial sun; for we know that the body of the sun gives out an intense white light, just as the lime-ball does, and that , like it, it has an atmosphere of glowing gases round it.
But perhaps we get the best idea of the mighty heat and light of the sun by remembering how few of the rays which dart out on all sides from this fiery ball can reach our tiny globe, and yet how powerful they are. Look at the globe of a lamp in the middle of the room, and see how its light pours out on all sides and into every corner; then take a grain of mustard-seed, which will very well represent the comparative size of our earth, and hold it up at a distance from the lamp. How very few of all those rays which are filling the room fall on the little mustard-seed, and just so few does our earth catch of the rays which dart out from the sun. And yet this small quantity (1/2000-millionth part of the whole) does nearly all the work of our world. (These and the preceding numerical statements will be found worked out in Sir J. Herschel's 'Familiar Lectures on Scientific Subjects,' 1868, from which many of the facts in

the first part of the lecture are taken.)
In order to see how powerful the sun's rays are, you have only to take a magnifying glass and gather them to a point on a piece of brown paper, for they will set the paper alight. Sir John Herschel tells us that at the Cape of Good Hope the heat was even so great that he cooked a beefsteak and roasted some eggs by merely putting them in the sun, in a box with a glass lid! Indeed, just as we should all be frozen to death if the sun were sold, so we should all be burnt up with intolerable heat if his fierce rays fell with all their might upon us. But we have an invisible veil protecting us, made - of what do you think? Of those tiny particles of water which the sunbeams draw up and scatter in the air, and which, as we shall see in Lecture IV, cut off part of the intense heat and make the air cool and pleasant for us.

Week 4
We have now learnt something of the distance, the size, the light, and the heat of the sun - the great source of the sunbeams. But we are as yet no nearer the answer to the question, What is a sunbeam? how does the sun touch our earth?
Now suppose I with to touch you from this platform where I stand, I can do it in two ways. Firstly, I can throw something at you and hit you - in this case a thing will have passed across the space from me to you. Or, secondly, if I could make a violent movement so as to shake the floor of the room, you would feel a quivering motion; and so I should touch you across the whole distance of the room. But in this case no thing would have passed from me to you but a movement or wave, which passed along the boards of the floor. Again, if I speak to you, how does the sound reach you ear? Not by anything being thrown from my mouth to your ear, but by the motion of the air. When I speak I agitate the air near my mouth, and that makes a wave in the air beyond, and that one, another, and another (as we shall see more fully in Lecture VI) till the last wave hits the drum of your ear.
Thus we see there are two ways of touching anything at a distance; 1st, by throwing some thing at it and hitting it; 2nd, by sending a movement of wave across to it, as in the case of the quivering boards and the air.
Now the great natural philosopher Newton thought that the sun touched us in the first of these ways, and that sunbeams were made of very minute atoms of matter thrown out by the sun, and making a perpetual cannonade on our eyes. It is easy to understand that this would make us see light and feel heat, just as a blow in the eye makes us see starts, or on

the body makes it feel hot: and for a long time this explanation was supposed to be the true one. But we know now that there are many facts which cannot be explained on this theory, though we cannot go into them here. What we will do, is to try and understand what now seems to be the true explanation of the sunbeam.

About the same time that Newton wrote, a Dutchman, named Huyghens, suggested that light comes from the sun in tiny waves, travelling across space much in the same way as ripples travel across a pond. The only difficulty was to explain in what substance these waves could be travelling: not through water, for we know that there is no water in space - nor through air, for the air stops at a comparatively short distance from our earth. There must then be something filling all space between us and the sun, finer than either water or air.

And now I must ask you to use all you imagination, for I want you to picture to yourselves something quite as invisible as the Emperor's new clothes in Andersen's fairy-tale, only with this difference, that our invisible something is very active; and though we can neither see it nor touch it we know it by its effects. You must imagine a fine substance filling all space between us and the sun and the starts. A substance so very delicate and subtle, that not only is it invisible, but it can pass through solid bodies such as glass, ice, or even wood or brick walls. This substance we call "ether." I cannot give you here the reasons why we must assume that it is throughout all space; you must take this on the word of such men as Sir John Herschel or Professor Clerk-Maxwell, until you can study the question for yourselves.

Now if you can imagine this ether filling every corner of space, so that it is everywhere and passes through everything, ask yourselves, what must happen when a great commotion is going on in one of the large bodies which float in it? When the atoms of the gases round the sun are clashing violently together to make all its light and heat, do you not think they must shake this ether all around them? And then, since the ether stretches on all sides from the sun to our earth and all other planets, must not this quivering travel to us, just as the quivering of the boards would from me to you? Take a basin of water to represent the ether, and take a piece of potassium like that which we used in our last lecture, and hold it with a pair of nippers in the middle of the water. You will see that as the potassium hisses and the flame burns round it, they will make waves which will travel all over the water to the edge of the basin,, and you can imagine how in the same way waves travel over the ether from the sun to us.

Straight away from the sun on all sides, never stopping, never resting, but chasing after each other with marvellous quickness, these tiny waves travel out into space by night and by day. When our spot of the earth where England lies is turned away from them and they cannot touch us, then it is night for us, but directly England is turned so as to face the sun, then they strike on the land, and the water, and warm it; or upon our eyes, making the nerves quiver so that we see light. Look up at the sun and picture to yourself that instead of one great blow from a fist causing you to see starts for a moment, millions of tiny blows from these sun-waves are striking every instant on you eye; then you will easily understand that his would cause you to see a constant blaze of light.

But when the sun is away, if the night is clear we have light from the starts. Do these then too make waves all across the enormous distance between them and us? Certainly they do, for they too are suns like our own, only they are so far off that the waves they send are more feeble, and so we only notice them when the sun's stronger waves are away.

But perhaps you will ask, if no one has ever seen these waves not the ether in which they are made, what right have we to say they are there? Strange as it may seem, though we cannot see them we have measured them and know how large they are, and how many can go into an inch of space. For as these tiny waves are running on straight forward through the room, if we put something in their way, they will have to run round it; and if you let in a very narrow ray of light through a shutter and put an upright wire in the sunbeam, you actually make the waves run round the wire just as water runs round a post in a river; and they meet behind the wire, just as the water meets in a V shape behind the post. Now when they meet, they run up against each other, and here it is we catch them. Fir if they meet comfortably, both rising up in a good wave, they run on together and make a bright line of light; but if they meet higgledy-piggledy, one up and the other down, all in confusion, they stop each other, and then there is no light but a line of darkness. And so behind your piece of wire you can catch the waves on a piece of paper, and you will find they make dark and light lines one side by side with the other, and by means of these bands it is possible to find out how large the waves must be. This question is too difficult for us to work it out here, but you can see that large waves will make broader light and dark bands than small ones will, and that in this way the size of the waves may be measured.

And now how large do you think they turn out to be? so very, very tiny that about fifty thousand waves are contained in a single inch of space! I have drawn on the board the length of an inch, and now I will measure the same space in the air between my finger and thumb. Within this space at this moment there are fifty thousand tiny waves moving up and down. I promised you we would find in science things as wonderful as in

fairy tales. Are not these tiny invisible messengers coming incessantly from the sun as wonderful as any fairies? and still more so when, as we shall see presently, they are doing nearly all the work of our world.
We must next try to realize how fast these waves travel. You will remember that an express train would take 171 years to reach us from the sun; and even a cannon-ball would take from ten to thirteen years to come that distance. Well, these tiny waves take only seven minutes and a half to come the whole 91 millions of miles. The waves which are hitting your eye at this moment are caused by a movement which began at the sun only 7 1/2 minutes ago. And remember, this movement is going on incessantly, and these waves are always following one after the other so rapidly that they keep up a perpetual cannonade upon the pupil of your eye. So fast do they come that about 608 billion waves enter your eye in one single second.* I do not ask you to remember these figures; I only ask you to try and picture to yourselves these infinitely tiny and active invisible messengers from the sun, and to acknowledge that light is a fairy thing. (*Light travels at the rate of 190,000 miles, or 12,165,120,000 inches in a second. Taking the average number of wave-lengths in an inch at 50,000, then 12,165,120,000 X 50,000 = 608,256,000,000,000.)
But we do not yet know all about our sunbeams. See, I have here a piece of glass with three sides, called a prism. If I put it in the sunlight which is streaming through the window, what happens? Look! on the table there is a line of beautiful colours. I can make it long or short, as I turn the prism, but the colours always remain arranged in the same way. Here at my left hand is the red, beyond it orange, then yellow, green, blue, indigo or deep blue, and violet, shading one into the other all along the line. We have all seen these colours dancing on the wall when the sun has been shining brightly on the cut-glass pendants of the chandelier, and you may see them still more distinctly if you let a ray of light into a darkened room, and pass it through the prism as in the diagram (Fig. 7). What are these colours? Do they come from the glass? No; for you will remember to have seen them in the rainbow, and in the soap- bubble, and even in a drop of dew or the scum on the top of a pond. This beautiful coloured line is only our sunbeam again, which has been split up into many colours by passing through the glass, as it is in the rain-drops of the rainbow and the bubbles of the scum of the pond.

Week 5

Till now we have talked of the sunbeam as if it were made of only one set of waves of different sizes, all travelling along together from the sun. These various waves have been measured, and we know that the waves which make up red light are larger and more lazy than those which make violet light, so that there are only thirty-nine thousand red waves in an inch, while there are fifty-seven thousand violet waves in the same space. How is it then, that if all these different waves making different colours, hit on our eye, they do not always make us see coloured light? Because, unless they are interfered with, they all travel along together, and you know that all colours, mixed together in proper proportion, make white.

I have here a round piece of cardboard, painted with the seven colours in succession several times over. When it is still you can distinguish them all apart, but when I whirl it quickly round - see! - the cardboard looks quite white, because we see them all so instantaneously that they are mingled together. In the same way light looks white to you, because all the different coloured waves strike on your eye at once. You can easily make on of these card for yourselves only the white will always look dirty, because you cannot get the colours pure.

Now, when the light passes through the three-sided glass or prism, the waves are spread out, and the slow, heavy, red waves lag behind and remain at the lower end R of the coloured line on the wall (Fig. 7), while the rapid little violet waves are bent more out of their road and run to V at the farther end of the line; and the orange, yellow, green, blue, and indigo arrange themselves between, according to the size of their waves.

And now you are very likely eager to ask why the quick waves should make us see one colour, and the slow waves another. This is a very difficult question, for we have a great deal still to learn about the effect of light on the eye. But you can easily imagine that colour is to our eye much the same as music is to our ear. You know we can distinguish different notes when the air-waves play slowly or quickly upon the drum of the ear (as we shall see in Lecture VI) and somewhat in the same way the tiny waves of the ether play on the retina or curtain at the back of our eye, and make the nerves carry different messages to the brain: and the colour we see depends upon the number of waves which play upon the retina in a second.

Do you think we have now rightly answered the question - What is a sunbeam? We have seen that it is really a succession of tiny rapid waves, travelling from the sun to us across the invisible substance we call "ether", and keeping up a constant cannonade upon everything which comes in their way. We have also seen that, tiny as these waves are, they can still vary in size, so that one single sunbeam is made up of myriads of different-sized waves, which travel all together and make us see white light; unless for some reason they are scattered apart, so that we see them separately as red, green, blue, or yellow. How they are scattered, and many other secrets of the sun-waves, we cannot stop to consider not, but must pass on to ask -

What work do the sunbeams do for us?

They do two things - they give us light and heat. It is by means of them alone that we see anything. When the room was dark you could not distinguish the table, the chairs, or even the walls of the room. Why? Because they had no light-waves to send to your eye. But as the sunbeams began to pour in at the window, the waves played upon the things in the room, and when they hit them they bounded off them back to your eye, as a wave of the sea bounds back from a rock and strikes against a passing boat. Then, when they fell upon your eye, they entered it and excited the retina and the nerves, and the image of the chair or the table was carried to your brain. Look around at all the things in this room. Is it not strange to think that each one of them is sending these invisible messengers straight to your eye as you look at it; and that you see me, and distinguish me from the table, entirely by the kind of waves we each send to you?

Some substances send back hardly any waves of light, but let them all pass through them, and thus we cannot see them. A pane of clear glass, for instance, lets nearly all the light-waves pass through it, and therefore you often cannot see that the glass is there, because no light-messengers come back to you from it. Thus people have sometimes walked up against a glass door and broken it, not seeing it was there. Those substances are transparent which, for some reason unknown to us, allow the ether waves to pass through them without shaking the atoms of which the substance is made. In clear glass, for example, all the light-waves pass through without affecting the substance of the glass; while in a white wall the larger part of the rays are reflected back to your eye, and those which pass into the wall, by giving motion to its atoms lose their own vibrations.

Into polished shining metal the waves hardly enter at all, but are thrown back from the surface; and so a steel knife or a silver spoon are very bright, and are clearly seen. Quicksilver is put at the back of looking-glasses because it reflects so many waves. It not only sends back those which come from the sun, but those, too, which come from your face. So, when you see yourself in a looking-glass, the sun-waves have first played on your face and bounded off from it to the looking-glass; then, when they strike the looking-glass, they are thrown back again on to the retina of your eye, and you see your own face by means of the very waves you threw off from it an instant before.

But the reflected light-waves do more for us than this. They not only make us see things, but they make us see them in different colours. What, you will ask, is this too the work of the sunbeams? Certainly; for if the colour we see depends on the size of the waves which come back to us, then we must see things coloured differently according to the waves they send back. For instance, imagine a sunbeam playing on a leaf: part of its waves bound straight back from it to our eye and make us see the surface of the leaf, but the rest go right into the leaf itself, and there some of them are used up and kept prisoners. The red, orange, yellow, blue, and violet waves are all useful to the leaf, and it does not let them go again. But it cannot absorb the green waves, and so it throws them back, and they travel to your eye and make you see a green colour. So when you say a leaf is green, you mean that the leaf does not want the green waves of the sunbeam, but sends them back to you. In the same way the scarlet geranium rejects the red waves; this table sends back brown waves; a white tablecloth sends back nearly the whole of the waves, and a black coat scarcely any. This is why, when there is very little light in the room, you can see a white tablecloth while you would not be able to distinguish a black object, because the few faint rays that are there, are all sent back to you from a white surface.

Is it not curious to think that there is really no such thing as colour in the leaf, the table, the coat, or the geranium flower, but we see them of different colours because, for some reason, they send back only certain coloured waves to our eye?

Wherever you look, then, and whatever you see, all the beautiful tints, colours, lights, and shades around you are the work of the tiny sun-waves.

Again, light does a great deal of work when it falls upon plants. Those rays of light which are caught by the leaf are by no means idle; we shall see in Lecture VII that the leaf uses them to digest its food and make the sap on which the plant feeds.

Week 6

We all know that a plant becomes pale and sickly if it has not sunlight, and the reason is, that without these light-waves it cannot get food out of the air, nor make the sap and juices which it needs. When you look at plants and trees growing in the beautiful meadows; at the fields of corn, and at the lovely landscape, you are looking on the work of the tiny waves of light, which never rest all through the day in helping to give life to every green thing that grows.

So far we have spoken only of light; but hold your hand in the sun and feel the heat of the sunbeams, and then consider if the waves of heat do not do work also. There are many waves in a sunbeam which move too slowly to make us see light when they hit our eye, but we can feel them as heat, though we cannot see them as light. The simplest way of feeling heat-waves is to hold a warm iron near your face. You know that no light comes from it, yet you can feel the heat-waves beating violently against your face and scorching it. Now there are many of these dark heat- rays in a sunbeam, and it is they which do most of the work in the world.

In the first place, as they come quivering to the earth, it is they which shake the water-drops apart, so that these are carried up in the air, as we shall see in the next lecture. And then remember, it is these drops, falling again as rain, which make the rivers and all the moving water on the earth. So also it is the heat-waves which make the air hot and light, and so cause it to rise and make winds and air-currents, and these again give rise to ocean-currents. It is these dark rays, again, which strike upon the land and give it the warmth which enables plants to grow. It is they also which keep up the warmth in our own bodies, both by coming to us directly from the sun, and also in a very roundabout way through plants. You will remember that plants use up rays of light and heat in growing; then either we eat the plants, or animals eat the plants and we eat the animals; and when we digest the food, that heat comes back in our bodies, which the plants first took from the sunbeam. Breathe upon your hand, and feel how hot your breath is; well, that heat which you feel, was once in a sunbeam, and has travelled from it through the food you have eaten, and has now been at work keeping up the heat of your body.

But there is still another way in which these plants may give out the heat-waves they have imprisoned. You will remember how we learnt in the first lecture that coal is made of plants, and that the heat they give out is the heat these plants once took in. Think how much work is done by burning coals. Not only are our houses warmed by coal fires and lighted by coal gas, but our steam-engines and machinery work entirely by water which has been turned into steam by the heat of coal and coke fire; and our steamboats travel all over the world by means of the same power. In the same way the oil of our lamps comes either from olives, which grow on trees; or from coal and the remains of plants and animals in the earth. Even our tallow candles are made of mutton fat, and sheep eat grass; as so, turn which way we will, we find that the light and heat on our earth, whether it comes from fires, or candles, or lamps, or gas, and whether it moves machinery, or drives a train, or propels a ship, is equally the work of the invisible waves of ether coming from the sun, which make what we call a sunbeam.

Lastly, there are still some hidden waves which we have not yet mentioned, which are not useful to us either as light or heat, and yet they are not idle.

Before I began this lecture, I put a piece of paper, which had been dipped in nitrate of silver, under a piece of glass; and between it and the glass I put a piece of lace. Look what the sun has been doing while I have been speaking. It has been breaking up the nitrate of silver on the paper and turning it into a deep brown substance; only where the threads of the lace were, and the sun could not touch the nitrate of silver, there the paper has remained light-coloured, and by this means I have a beautiful impression of the lace on the paper. I will now dip the impression into water in which some hyposulphite of soda is dissolved, and this will "fix" the picture, that is, prevent the sun acting upon it any more; then the picture will remain distinct, and I can pass it round to you all. Here, again, invisible waves have been at work, and this time neither as light nor as heat, but as chemical agents, and it is these waves which give us all our beautiful photographs. In any toyshop you can buy this prepared paper, and set the chemical waves at work to make pictures. Only you must remember to fix it in the solution afterwards, otherwise the chemical rays will go on working after you have taken the lace away, and all the paper will become brown and your picture will disappear.

And now, tell me, may we not honestly say, that the invisible waves which make our sunbeams, are wonderful fairy messengers as they travel eternally and unceasingly across space, never resting, never tiring in doing the work of our world? Little as we have been able to learn about them in one short hour, do they not seem to you worth studying and worth thinking about, as we look at the beautiful results of their work? The ancient Greeks worshipped the sun, and condemned to death one of their greatest philosophers, named Anaxagoras, because he denied that it was a god. We can scarcely wonder at this when we see what the sun does for our world; but we know that it is a huge globe made of gases and fiery matter and not a god. We are grateful for the sun instead of to him, and surely we shall look at him with new interest, now that we can picture his tiny messengers, the sunbeams, flitting over all space, falling upon our earth, giving us light to see with, and beautiful colours to enjoy, warming the air and the earth, making the refreshing rain, and, in a word, filling the world with life and gladness.

Week 7

LECTURE III The Aerial Ocean in Which We Live

Did you ever sit on the bank of a river in some quiet spot where the water was deep and clear, and watch the fishes swimming lazily along? When I was a child this was one of my favourite occupations in the summertime on the banks of the Thames, and there was one question which often puzzled me greatly, as I watched the minnows and gudgeon gliding along through the water. Why should fishes live in something and be often buffeted about by waves and currents, while I and others lived on the top of the earth and not in anything? I do not remember ever asking anyone about this; and if I had, in those days people did not pay much attention to children's questions, and probably nobody would have told me, what I now tell you, that we do live in something quite as real and often quite as rough and stormy as the water in which the fishes swim. The something in which we live is air, and the reason that we do not perceive it, is that we are in it, and that it is a gas, and invisible to us; while we are above the water in which the fishes live, and it is a liquid which our eyes can perceive.

But let us suppose for a moment that a being, whose eyes were so made that he could see gases as we see liquids, was looking down from a distance upon our earth. He would see an ocean of air, or aerial ocean, all round the globe, with birds floating about in it, and people walking along the bottom, just as we see fish gliding along the bottom of a river. It is true, he would never see even the birds come near to the surface, for the highest- flying bird, the condor, never soars more than five miles from the ground, and our atmosphere, as we shall see, is at least 100 miles high. So he would call us all deep-air creatures, just as we talk of deep-sea animals; and if we can imagine that he fished in this air-ocean, and could pull one of us out of it into space, he would find that we should gasp and die just as fishes do when pulled out of the water.

He would also observe very curious things going on in our air- ocean; he would see large streams and currents of air, which we call winds, and which would appear to him as ocean-currents do to us, while near down to the earth he would see thick mists forming and then disappearing again, and these would be our clouds. From them he would see rain, hail and snow falling to the earth, and from time to time bright flashes would shoot across the air- ocean, which would be our lightning. Nay even the brilliant rainbow, the northern aurora borealis, and the falling stars, which seem to us so high up in space, would be seen by him near to our earth, and all within the aerial ocean.

But as we know of no such being living in space, who can tell us what takes place in our invisible air, and we cannot see it ourselves, we must try by experiments to see it with our imagination, though we cannot with our eyes.

First, then, can we discover what air is? At one time it was thought that it was a simple gas and could not be separated into more than one kind. But we are now going to make an experiment by which it has been shown that air is made of two gases mingled together, and that one of these gases, called oxygen, is used up when anything burns, while the other nitrogen is not used, and only serves to dilute the minute atoms of oxygen. I have here a glass bell-jar, with a cork fixed tightly in the neck, and I place the jar over a pan of water, while on the water floats a plate with a small piece of phosphorus upon it. You will see that by putting the bell-jar over the water, I have shut in a certain quantity of air, and my object now is to use up the oxygen out of this air and leave only nitrogen behind. To do this I must light the piece of phosphorus, for you will remember it is in burning that oxygen is used up. I will take the cork out, light the phosphorus, and cork up the jar again. See! as the phosphorus burns white fumes fill the jar. These fumes are phosphoric acid which is a substance made of phosphorous and the oxygen of the air together.

Now, phosphoric acid melts in water just as sugar does, and in a few minutes these fumes will disappear. They are beginning to melt already, and the water from the pan is rising up in the bell-jar. Why is this? Consider for a moment what we have done. First, the jar was full of air, that is, of mixed oxygen and nitrogen; then the phosphorus used up the oxygen making white fumes; afterwards, the water sucked up these fumes; and so, in the jar now nitrogen is the only gas left, and the water has risen up to fill all the rest of the space that was once taken up with oxygen.

We can easily prove that there is no oxygen now in the jar. I take out the cork and let a lighted taper down into the gas. If there were any oxygen the taper would burn, but you see it goes out directly proving that all the oxygen has been used up by the phosphorous. When this experiment is made very accurately, we find that for every pint of oxygen in air there are four pints of nitrogen, so that the active oxygen-atoms are scattered about, floating in the sleepy, inactive nitrogen.

It is these oxygen-atoms which we use up when we breathe. If I had put a mouse under the bell-jar, instead of the phosphorus, the water would have risen just the same, because the mouse would have breathed in the oxygen and used it up in its body, joining it to carbon and making a bad gas, carbonic acid, which would also melt in the water, and when all the oxygen was used, the mouse would have died.

Do you see now how foolish it is to live in rooms that are closely shut up, or to hide your head under the bedclothes when you sleep? You use up all the oxygen-atoms, and then there are none left for you to breathe; and besides this, you send out of your mouth bad fumes, though you cannot see them, and these, when you breathe them in again, poison you and

make you ill.
Perhaps you will say, If oxygen is so useful, why is not the air made entirely of it? But think for a moment. If there was such an immense quantity of oxygen, how fearfully fast everything would burn! Our bodies would soon rise above fever heat from the quantity of oxygen we should take in, and all fires and lights would burn furiously. In fact, a flame once lighted would spread so rapidly that no power on earth could stop it, and everything would be destroyed. So the lazy nitrogen is very useful in keeping the oxygen-atoms apart; and we have time, even when a fire is very large and powerful, to put it out before it has drawn in more and more oxygen from the surrounding air. Often, if you can shut a fire into a closed space, as in a closely-shut room or the hold of a ship, it will go out, because it has used up all the oxygen in the air.
So, you see, we shall be right in picturing this invisible air all around us as a mixture of two gases. But when we examine ordinary air very carefully, we find small quantities of other gases in it, besides oxygen and nitrogen. First, there is carbonic acid gas. This is the bad gas which we give out of our mouths after we have burnt up the oxygen with the carbon of our bodies inside our lungs; and this carbonic acid is also given out from everything that burns. If only animals lived in the world, this gas would soon poison the air; but plants get hold of it, and in the sunshine they break it up again, as we shall see in Lecture VII, and use up the carbon, throwing the oxygen back into the air for us to use. Secondly, there are very small quantities of ammonia, or the gas which almost chokes you in smelling-salts, and which, when liquid is commonly called "spirits of hartshorn." This ammonia is useful to plants, as we shall see by and by. Lastly, there is a great deal of water in the air, floating about as invisible vapour or water-dust, and this we shall speak of in the next lecture. Still, all these gases and vapours in the atmosphere are in very small quantities, and the bulk of the air is composed of oxygen and nitrogen.
Having now learned what air is, the next question which presents itself is, Why does it stay round our earth? You will remember we saw in the first lecture, that all the little atoms of a gas are trying to fly away from each other, so that if I turn on this gas-jet the atoms soon leave it, and reach you at the farther end of the room, and you can smell the gas. Why, then, do not all the atoms of oxygen and nitrogen fly away from our earth into space, and leave us without any air?
Ah! here you must look for another of our invisible forces. Have you forgotten our giant force, "gravitation," which draws things together from a distance? This force draws together the earth and the atoms of oxygen and nitrogen; and as the earth is very big and heavy, and the

atoms of air are light and easily moved, they are drawn down to the earth and held there by gravitation. But for all that, the atmosphere does not leave off trying to fly away; it is always pressing upwards and outwards with all its might, while the earth is doing its best to hold it down.

The effect of this is, that near the earth, where the pull downward is very strong, the air-atoms are drawn very closely together, because gravitation gets the best of the struggle. But as we get farther and farther from the earth, the pull downward becomes weaker, and then the air-atoms spring farther apart, and the air becomes thinner. Suppose that the lines in this diagram represent layers of air. Near the earth we have to represent them as lying closely together, but as they recede from the earth they are also farther apart.

But the chief reason why the air is thicker or denser nearer the earth, is because the upper layers press it down. If you have a heap of papers lying one on the top of the other, you know that those at the bottom of the heap will be more closely pressed together than those above, and just the same is the case with the atoms of the air. Only there is this difference, if the papers have lain for some time, when you take the top ones off, the under ones remain close together. But it is not so with the air, because air is elastic, and the atoms are always trying to fly apart, so that directly you take away the pressure they spring up again as far as they can.

Week 8

I have here an ordinary pop-gun. If I push the cork in very tight, and then force the piston slowly inwards, I can compress the air a good deal. Now I am forcing the atoms nearer and nearer together, but at last they rebel so strongly against being more crowded that the cork cannot resist their pressure. Out it flies, and the atoms spread themselves out comfortably again in the air all around them. Now, just as I pressed the air together in the pop-gun, so the atmosphere high up above the earth presses on the air below and keeps the atoms closely packed together. And in this case the atoms cannot force back the air above them as they did the cork in the pop-gun; they are obliged to submit to be pressed together.

Even a short distance from the earth, however, at the top of a high mountain, the air becomes lighter, because it has less weight of atmosphere above it, and people who go up in balloons often have great difficulty in breathing, because the air is so thin and light. In 1804 a Frenchman, named Gay-Lussac, went up four miles and a half in a balloon, and brought down some air; and he found that it was much less heavy than the same quantity of air taken close down to the earth, showing that it was much thinner, or rarer, as it is called;* and when, in 1862, Mr. Glaisher and Mr. Coxwell went up five miles and a half, Mr. Glaisher's veins began to swell, and his head grew dizzy, and he fainted. The air was too thin for him to breathe enough in at a time, and it did not press heavily enough on the drums of his ears and the veins of his body. He would have died if Mr. Coxwell had not quickly let off some of the gas in the balloon, so that it sank down into denser air. (*100 cubic inches near the earth weighed 31 grains, while the same quantity taken at four and a half miles up in the air weighed only 12 grains, or two- fifths of the weight.)

And now comes another very interesting question. If the air gets less and less dense as it is farther from the earth, where does it stop altogether? We cannot go up to find out, because we should die long before we reached the limit; and for a long time we had to guess about how high the atmosphere probably was, and it was generally supposed not to be more than fifty miles. But lately, some curious bodies, which we should have never suspected would be useful to us in this way, have let us into the secret of the height of the atmosphere. These bodies are the meteors, or falling stars.

Most people, at one time or another, have seen what looks like a star shoot right across the sky, and disappear. On a clear starlight night you may often see one or more of these bright lights flash through the air; for one falls on an average in every twenty minutes, and on the nights of August 9th and November 13th there are numbers in one part of the sky. These bodies are not really stars; they are simply stones or lumps of metal flying through the air, and taking fire by clashing against the atoms of oxygen in it. There are great numbers of these masses moving round and round the sun, and when our earth comes across their path, as it does especially in August and November, they dash with such tremendous force through the atmosphere that they grow white-hot, and give out light, and then disappear, melted into vapour. Every now and then one falls to the earth before it is all melted away, and thus we learn that these stones contain tin, iron, sulphur, phosphorus, and other substances.

It is while these bodies are burning that they look to us like falling stars, and when we see them we know that hey must be dashing against our atmosphere. Now if two people stand a certain known distance, say fifty miles, apart on the earth and observe these meteors and the direction in which they each see them fall, they can calculate (by means of the angle between the two directions) how high they are above them when they first see them, and at that moment they must have struck against the atmosphere, and even travelled some way through it, to become white-hot. In this way we have learnt that meteors burst into light at least 100 miles above the surface of the earth, and so the atmosphere must be more than 100 miles high.

Our next question is as to the weight of our aerial ocean. You will easily understand that all this air weighing down upon the earth must be very heavy, even though it grows lighter as it ascends. The atmosphere does, in fact, weigh down upon land at the level of the sea as much as if a 15-pound weight were put upon every square inch of land. This little piece of linen paper, which I am holding up, measures exactly a square inch, and as it lies on the table, it is bearing a weight of 15 lbs. on its surface. But how, then, comes it that I can lift it so easily? Why am I not conscious of the weight?

To understand this you must give all your attention, for it is important and at first not very easy to grasp. you must remember, in the first place, that the air is heavy because it is attracted to the earth, and in the second place, that since air is elastic all the atoms of it are pushing upwards against this gravitation. And so, at any point in air, as for instance the place where the paper now is as I hold it up, I feel no pressure because exactly as much as gravitation is pulling the air down, so much elasticity is resisting and pushing it up. So the pressure is equal upwards, downwards, and on all sides, and I can move the paper with equal ease any way.

Even if I lay the paper on the table this is still true, because there is always some air under it. If, however, I could get the air quite away from one side of the paper, then the pressure on the other side would show itself. I can do this by simply wetting the paper and letting it fall on the table, and the water will prevent any air from getting under it. Now see! if I try to lift it by the thread in the middle, I have great difficulty, because the whole 15 pounds' weight of the atmosphere is pressing it down. A still better way of making the experiment is with a piece of leather, such as the boys often amuse themselves with in the streets. This piece of leather has been well soaked. I drop it on the floor and see! it requires all my strength to pull it up. (In fastening the string to the leather the hole must be very small and the know as flat as possible, and it is even well to put a small piece of kid under the knot. When I first made this experiment, not having taken these precautions, it did not succeed well, owing to air getting in through the hole.) I now drop it on this stone weight, and so heavily is it pressed down upon it by the atmosphere that I can lift the weight without its breaking away from it.

Have you ever tried to pick limpets off a rock? If so, you know how tight they cling. the limpet clings to the rock just in the same way as this leather does to the stone; the little animal exhausts the air inside it's shell, and then it is pressed against the rock by the whole weight of the air above.

Perhaps you will wonder how it is that if we have a weight of 15 lbs. pressing on every square inch of our bodies, it does not crush us. And, indeed, it amounts on the whole to a weight of about 15 tons upon the body of a grown man. It would crush us if it were not that there are gases and fluids inside our bodies which press outwards and balance the weight so that we do not feel it at all.

This is why Mr. Glaisher's veins swelled and he grew giddy in thin air. The gases and fluids inside his body were pressing outwards as much as when he was below, but the air outside did not press so heavily, and so all the natural condition of his body was disturbed.

I hope we now realize how heavily the air presses down upon our earth, but it is equally necessary to understand how, being elastic, it also presses upwards; and we can prove this by a simple experiment. I fill this tumbler with water, and keeping a piece of card firmly pressed against it, I turn the whole upside- down. When I now take my hand away you would naturally expect the card to fall, and the water to be spilt. But no! the card remains as if glued to the tumbler, kept there entirely by the air pressing upwards against it. (The engraver has drawn the tumbler only half full of water. The experiment will succeed quite as well in this way if the tumbler be turned over quickly, so that part of the air escapes between the tumbler and the card, and therefore the space above the water is occupied by air less dense than that outside.)

And now we are almost prepared to understand how we can weigh the invisible air. One more experiment first. I have here what is called a U tube, because it is shaped like a large U. I pour some water in it till it is about half full, and you will notice that the water stands at the same height in both arms of the tube, because the air presses on both surfaces alike. Putting my thumb on one end I tilt the tube carefully, so as to make the water run up to the end of one arm, and then turn it back again. But the water does not now return to its even position, it remains up in the arm on which my thumb rests. Why is this? Because my thumb keeps back the air from pressing at that end, and the whole weight of the atmosphere rests on the water at the other end. And so we learn that not only has the atmosphere real weight, but we can see the effects of this weight by making it balance a column of water or any other liquid. In the case of the wetted leather we felt the weight of the air, here we see its effects.

Now when we wish to see the weight of the air we consult a barometer, which works really just in the same way as the water in this tube. An ordinary upright barometer is simply a straight tube of glass filled with mercury or quicksilver, and turned upside-down in a small cup of mercury. The tube is a little more than 30 inches long, and though it is quite full of mercury before it is turned up, yet directly it stands in the cup the mercury falls, till there is a height of about 30 inches between the surface of the mercury in the cup, and that of the mercury in the tube. As it falls it leaves an empty space above the mercury which is called a vacuum, because it has no air in it. Now, the mercury is under the same conditions as the water was in the U tube, there is no pressure upon it at the top of the tube, while there is a pressure of 15 lbs. upon it in the bowl, and therefore it remains held up in the tube.

Week 9

But why will it not remain more than 30 inches high in the tube? You must remember it is only kept up in the tube at all by the air which presses on the mercury in the cup. And that column of mercury now balances the pressure of the air outside, and presses down on the mercury in the cup at its mouth just as much as the air does on the rest. So this cup and tube act exactly like a pair of scales. The air outside is the thing to be weighed at one end as it presses on the mercury, the column answers to the leaden weight at the other end which tells you how heavy the air is. Now if the bore of this tube is made an inch square, then the 30 inches of mercury in it weigh exactly 15 lbs, and so we know that the weight of the air is 15 lbs. upon every square inch, but if the bore of the tube is only half a square inch, and therefore the 30 inches of mercury only weigh 7 1/2 lbs. instead of 15 lbs., the pressure of the atmosphere will also be halved, because it will only act upon half a square inch of surface, and for this reason it will make no difference to the height of the mercury whether the tube be broad or narrow.

But now suppose the atmosphere grows lighter, as it does when it has much damp in it. The barometer will show this at once, because there will be less weight on the mercury in the cup, therefore it will not keep the mercury pushed so high up in the tube. In other words, the mercury in the tube will fall.

Let us suppose that one day the air is so much lighter that it presses down only with a weight of 14 1/2 lbs. to the square inch instead of 15 lbs. Then the mercury would fall to 29 inches, because each inch is equal to the weight of half a pound. Now, when the air is damp and very full of water-vapour it is much lighter, and so when the barometer falls we expect rain. Sometimes, however, other causes make the air light, and

then, although the barometer is low, no rain comes,
Again, if the air becomes heavier the mercury is pushed up above 30 to 31 inches, and in this way we are able to weigh the invisible air-ocean all over the world, and tell when it grows lighter or heavier. This then, is the secret of the barometer. We cannot speak of the thermometer today, but I should like to warn you in passing that it has nothing to do with the weight of the air, but only with heat, and acts in quite a different way.
And now we have been so long hunting out, testing and weighing our aerial ocean, that scarcely any time is left us to speak of its movements or the pleasant breezes which it makes for us in our country walks. Did you ever try to run races on a very windy day? Ah! then you feel the air strongly enough; how it beats against your face and chest, and blows down your throat so as to take your breath away; and what hard work it is to struggle against it! Stop for a moment and rest, and ask yourself, what is the wind? Why does it blow sometimes one way and sometimes another, and sometimes not at all?
Wind is nothing more than air moving across the surface of the earth, which as it passes along bends the tops of the trees, beats against the houses, pushes the ships along by their sails, turns the windmill, carries off the smoke from cities, whistles through the keyhole, and moans as it rushes down the valley. What makes the air restless? why should it not lie still all round the earth?
It is restless because, as you will remember, its atoms are kept pressed together near the earth by the weight of the air above, and they take every opportunity, when they can find more room, to spread out violently and rush into the vacant space, and this rush we call a wind.
Imagine a great number of active schoolboys all crowded into a room till they can scarcely move their arms and legs for the crush, and then suppose all at once a large door is opened. Will they not all come tumbling out pell-mell, one over the other, into the hall beyond, so that if you stood in their way you would most likely be knocked down? Well, just this happens to the air- atoms; when they find a space before them into which they can rush, they come on helter-skelter, with such force that you have great difficulty in standing against them, and catch hold of something to support you for fear you should be blown down.
But how come they to find any empty space to receive them? To answer this we must go back again to our little active invisible fairies the sunbeams. When the sun-waves come pouring down upon the earth they pass through the air almost without heating it. But not so with the ground; there they pass down only a short distance and then are thrown back again. And when these sun- waves come quivering back they force the atoms of the air near the earth apart and make it lighter; so that the air

close to the surface of the heated ground becomes less heavy than the air above it, and rises just as a cork rises in water. You know that hot air rises in the chimney; for if you put a piece of lighted paper on the fire it is carried up by the draught of air, often even before it can ignite. Now just as the hot air rises from the fire, so it rises from the heated ground up into higher parts of the atmosphere. and as it rises it leaves only thin air behind it, and this cannot resist the strong cold air whose atoms are struggling and trying to get free, and they rush in and fill the space.

One of the simplest examples of wind is to be found at the seaside. there in the daytime the land gets hot under the sunshine, and heats the air, making it grow light and rise. Meanwhile the sunshine on the water goes down deeper, and so does not send back so many heat-waves into the air; consequently the air on the top of the water is cooler and heavier, and it rushes in from over the sea to fill up the space on the shore left by the warm air as it rises. This is why the seaside is so pleasant in hot weather. During the daytime a light sea-breeze nearly always sets in from the sea to the land.

When night comes, however, then the land loses its heat very quickly, because it has not stored it up and the land-air grows cold; but the sea, which has been hoarding the sun-waves down in its depths, now gives them up to the atmosphere above it, and the sea-air becomes warm and rises. For this reason it is now the turn of the cold air from the land to spread over the sea, and you have a land-breeze blowing off the shore.

Again, the reason why there are such steady winds, called the trade winds, blowing towards the equator, is that the sun is very hot at the equator, and hot air is always rising there and making room for colder air to rush in. We have not time to travel farther with the moving air, though its journeys are extremely interesting; but if, when you read about the trade and other winds, you will always picture to yourselves warm air made light by the heat rising up into space and cold air expanding and rushing in to fill its place, I can promise you that you will not find the study of aerial currents so dry as many people imagine it to be.

We are now able to form some picture of our aerial ocean. We can imagine the active atoms of oxygen floating in the sluggish nitrogen, and being used up in every candle-flame, gas-jet and fire, and in the breath of all living beings; and coming out again tied fast to atoms of carbon and making carbonic acid. Then we can turn to trees and plants, and see them tearing these two apart again, holding the carbon fast and sending the invisible atoms of oxygen bounding back again into the air, ready to recommence work. We can picture all these air-atoms, whether of oxygen or nitrogen, packed close together on the surface of the earth, and lying gradually farther and farther apart, as they have less weight above them, till they become so scattered that we can only detect them as they rub against the flying meteors which flash into light. We can feel this great weight of air pressing the limpet on to the rock; and we can see it pressing up the mercury in the barometer and so enabling us to measure its weight. Lastly, every breath of wind that blows past us tells us how this aerial ocean is always moving to and fro on the face of the earth; and if we think for a moment how much bad air and bad matter it must carry away, as it goes from crowded cities to be purified in the country, we can see how, in even this one way alone, it is a great blessing to us.

Yet even now we have not mentioned many of the beauties of our atmosphere. It is the tiny particles floating in the air which scatter the light of the sun so that it spreads over the whole country and into shady places. The sun's rays always travel straight forward; and in the moon, where there is no atmosphere, there is no light anywhere except just where the rays fall. But on our earth the sun-waves hit against the myriads of particles in the air and glide off them into the corners of the room or the recesses of a shady lane, and so we have light spread before us wherever we walk in the daytime, instead of those deep black shadows which we can see through a telescope on the face of the moon.

Again, it is electricity playing in the air-atoms which gives us the beautiful lightning and the grand aurora borealis, and even the twinkling of the starts is produced entirely by minute changes in the air. If it were not for our aerial ocean, the stars would stare at us sternly, instead of smiling with the pleasant twinkle-twinkle which we have all learned to love as little children.

All these questions, however, we must leave for the present; only I hope you will be eager to read about them wherever you can, and open your eyes to learn their secrets. For the present we must be content if we can even picture this wonderful ocean of gas spread round our earth, and some of the work it does for us.

We said in the last lecture that without the sunbeams the earth would be cold, dark, and frost-ridden. With sunbeams, but without air, it would indeed have burning heat, side by side with darkness and ice, but it could have no soft light. our planet might look beautiful to others, as the moon does to us, but it could have comparatively few beauties of its own. With the sunbeams and the air, we see it has much to make it beautiful. But a third worker is wanted before our planet can revel in activity and life. This worker is water; and in the next lecture we shall learn something of the beauty and the usefulness of the "drops of water" on their travels.

Week 10

LECTURE IV. A DROP OF WATER ON ITS TRAVELS

We are going to spend an hour to-day in following a drop of water on its travels. If I dip my finger in this basin of water and lift it up again, I bring with it a small glistening drop out of the body of water below, and hold it before you. Tell me, have you any idea where this drop has been? what changes it has undergone, and what work it has been doing during all the long ages that water has lain on the face of the earth? It is a drop now, but it was not so before I lifted it out of the basin; then it was part of a sheet of water, and will be so again if I let it fall. Again, if I were to put this basin on the stove till all the water had boiled away, where would my drop be then? Where would it go? What forms will it take before it reappears in the rain-cloud, the river, or the sparkling dew?

These are questions we are going to try to answer to-day; and first, before we can in the least understand how water travels, we must call to mind what we have learnt about the sunbeams and the air. We must have clearly pictured in our imagination those countless sun-waves which are for ever crossing space, and especially those larger and slower undulations, the dark heat- waves; for it is these, you will remember, which force the air- atoms apart and make the air light, and it is also these which are most busy in sending water on its travels. But not these alone. The sun-waves might shake the water-drops as much as they liked and turn them into invisible vapour, but they could not carry them over the earth if it were not for the winds and currents of that aerial ocean which bears the vapour on its bosom, and wafts it to different regions of the world.

Let us try to understand how these two invisible workers, the sun-waves and the air, deal with the drops of water. I have here a kettle (Fig. 18, p. 76) boiling over a spirit-lamp, and I want you to follow minutely what is going on in it. First, in the flame of the lamp, atoms of the spirit drawn up from below are clashing with the oxygen-atoms in the air. This, as you know, causes heat-waves and light-waves to move rapidly all round the lamp. The light-waves cannot pass through the kettle, but the heat-waves can, and as they enter the water inside they agitate it violently. Quicker, and still more quickly, the particles of water near the bottom of the kettle move to and fro and are shaken apart; and as they become light they rise through the colder water letting another layer come down to be heated in its turn. The motion grows more and more violent, making the water hotter and hotter, till at last the particles of which it is composed fly asunder, and escape as invisible vapour. If this kettle were transparent you would not see any steam above the water, because it is in the form of an invisible gas. But as the steam comes out of the mouth of the kettle you see a cloud. Why is this? Because the vapour is chilled by coming out into the cold air, and its particles are drawn together again into tiny, tiny drops of water, to which Dr. Tyndall has given the suggestive name of water-dust. If you hold a plate over the steam you can catch these tiny drops, though they will run into one another almost as you are catching them.

The clouds you see floating in the sky are made of exactly the same kind of water-dust as the cloud from the kettle, and I wish to show you that this is also really the same as the invisible steam within the kettle. I will do so by an experiment suggested by Dr. Tyndall. Here is another spirit-lamp, which I will hold under the cloud of steam - see! the cloud disappears! As soon as the water-dust is heated the heat-waves scatter it again into invisible particles, which float away into the room. Even without the spirit-lamp, you can convince yourself that water-vapour may be invisible; for close to the mouth of the kettle you will see a short blank space before the cloud begins. In this space there must be steam, but it is still so hot that you cannot see it; and this proves that heat-waves can so shake water apart as to carry it away invisibly right before your eyes.

Now, although we never see any water travelling from our earth up into the skies, we know that it goes there, for it comes down again in rain, and so it must go up invisibly. But where does the heat come from which makes this water invisible? Not from below, as in the case of the kettle, but from above, pouring down from the sun. Wherever the sun-waves touch the rivers, ponds, lakes, seas, or fields of ice and snow upon our earth, they carry off invisible water-vapour. They dart down through the top layers of the water, and shake the water-particles forcibly apart; and in this case the drops fly asunder more easily and before they are so hot, because they are not kept down by a great weight of water above, as in the kettle, but find plenty of room to spread themselves out in the gaps between the air-atoms of the atmosphere.

Can you imagine these water-particles, just above any pond or lake, rising up and getting entangled among the air-atoms? They are very light, much lighter than the atmosphere; and so, when a great many of them are spread about in the air which lies just over the pond, they make it much lighter than the layer of air above, and so help it to rise, while the heavier layer of air comes down ready to take up more vapour.

In this way the sun-waves and the air carry off water everyday, and all day long, from the top of lakes, rivers, pools, springs, and seas, and even from the surface of ice and snow. Without any fuss or noise or sign of any kind, the water of our earth is being drawn up invisibly into the sky.

It has been calculated that in the Indian Ocean three-quarters of an inch of water is carried off from the surface of the sea in one day and night; so that as much as 22 feet, or a depth of water about twice the height of an ordinary room, is silently and invisibly lifted up from the whole surface of the ocean in one year. It is true this is one of the hottest parts of the earth, where the sun-waves are most active; but even in our own country many feet of water are drawn up in the summer-time.

What, then, becomes of all this water? Let us follow it as it struggles upwards to the sky. We see it in our imagination first carrying layer after layer of air up with it from the sea till it rises far above our heads and above the highest mountains. But now, call to mind what happens to the air as it recedes from the earth. Do you not remember that the air-atoms are always trying to fly apart, and are only kept pressed together by the weight of air above them? Well, so this water-laden air rises up, its particles, no longer so much pressed together, begin to separate, and as all work requires an expenditure of heat, the air becomes colder, and then you know at once what must happen to the invisible vapour, — it will form into tiny water-drops, like the steam from the kettle. And so, as the air rises and becomes colder, the vapour gathers into the visible masses, and we can see it hanging in the sky, and call it clouds. When these clouds are highest they are about ten miles from the earth, but when they are made of heavy drops and hang low down, they sometimes come within a mile of the ground.

Look up at the clouds as you go home, and think that the water of which they are made has all been drawn up invisibly through the air. Not, however, necessarily here in London, for we have already seen that air travels as wind all over the world, rushing in to fill spaces made by rising air wherever they occur, and so these clouds may be made of vapour collected in the Mediterranean, or in the Gulf of Mexico off the coast of America, or even, if the wind is from the north, of chilly particles gathered from the surface of Greenland ice and snow, and brought here by the moving currents of air. Only, of one thing we may be sure, that they come from the water of our earth.

Sometimes, if the air is warm, these water-particles may travel a long way without ever forming into clouds; and on a hot, cloudless day the air is often very full of invisible vapour. Then, if a cold wind comes sweeping along, high up in the sky, and chills this vapour, it forms into great bodies of water-dust clouds, and the sky is overcast. At other times clouds hang lazily in a bright sky, and these show us that just where they are (as in Fig. 19) the air is cold and turns the invisible vapour rising from the ground into visible water-dust, so that exactly in those spaces we see it as clouds. Such clouds form often on warm, still summer's day, and they are shaped like masses of wool, ending in a straight line below. They are not merely hanging in the sky, they are really resting upon a tall column of invisible vapour which stretches right up from the earth; and that straight line under the clouds marks the place where the air becomes cold enough to turn this invisible vapour into visible drops of water.

Week 11

And now, suppose that while these or any other kind of clouds are overhead, there comes along either a very cold wind, or a wind full of vapour. As it passes through the clouds, it makes them very full of water, for, if it chills them, it makes the water- dust draw more closely together; or, if it brings a new load of water-dust, the air is fuller than it can hold. In either case a number of water-particles are set free, and our fairy force "cohesion" seizes upon them at once and forms them into large water-drops. Then they are much heavier than the air, and so they can float no longer, but down they come to the earth in a shower of rain.

There are other ways in which the air may be chilled, and rain made to fall, as, for example, when a wind laden with moisture strikes against the cold tops of mountains. Thus the Khasia Hills in India which face the Bay of Bengal, chill the air which crosses them on its way from the Indian Ocean. The wet winds are driven up the sides of the hills, the air expands, and the vapour is chilled, and forming into drops, falls in torrents of rain. Sir J. Hooker tells us that as much as 500 inches of rain fell in these hills in nine months. That is to say, if you could measure off all the ground over which the rain fell, and spread the whole nine months' rain over it, it would make a lake 500 inches, or more than 40 feet deep! You will not be surprised that the country on the other side of these hills gets hardly any rain, for all the water has been taken out of the air before it comes there. Again for example in England, the wind comes to Cumberland and Westmorland over the Atlantic, full of vapour, and as it strikes against the Pennine Hills it shakes off its watery load; so that the lake district is the most rainy in England, with the exception perhaps of Wales, where the high mountains have the same effect.

In this way, from different causes, the water of which the sun has robbed our rivers and seas, comes back to us, after it has travelled to various parts of the world, floating on the bosom of the air. But it does not always fall straight back into the rivers and seas again, a large part of it falls on the land, and has to trickle down slopes and into the earth, in order to get back to its natural home, and it is often caught on its way before it can reach the great waters.

Go to any piece of ground which is left wild and untouched you will find it covered with grass weeds, and other plants; if you dig up a small plot you will find innumerable tiny roots creeping through the ground in every direction. Each of these roots has a sponge-like mouth by which the plant takes up water. Now, imagine rain-drops falling on this plot of ground and sinking into the earth. On every side they will find rootlets thirsting to drink them in, and they will be sucked up as if by tiny sponges, and drawn into the plants, and up the stems to the leaves. Here, as we shall see in Lecture VII., they are worked up into food for the plant, and only if the leaf has more water than it needs, some drops may escape at the tiny openings under the leaf, and be drawn up again by the sun-waves as invisible vapour into the air.

Again, much of the rain falls on hard rock and stone, where it cannot sink in, and then it lies in pools till it is shaken apart again into vapour and carried off in the air. Nor is it idle here, even before it is carried up to make clouds. We have to thank this invisible vapour in the air for protecting us from the burning heat of the sun by day and intolerable frost by night.

Let us for a moment imagine that we can see all that we know exists between us and the sun. First, we have the fine ether across which the sunbeams travel, beating down upon our earth with immense force, so that in the sandy desert they are like a burning fire. Then we have the coarser atmosphere of oxygen and nitrogen atoms hanging in this ether, and bending the minute sun- waves out of their direct path. But they do very little to hinder them on their way, and this is why in very dry countries the sun's heat is so intense. The rays beat down mercilessly, and nothing opposes them. Lastly, in damp countries we have the larger but still invisible particles of vapour hanging about among the air-atoms. Now, these watery particles, although they are very few (only about one twenty-fifth part of the whole atmosphere), do hinder the sun-waves. For they are very greedy of heat, and though the light-waves pass easily through them, they catch the heat-waves and use them to help themselves to expand. And so, when there is invisible vapour in the air, the sunbeams come to us deprived of some of their heat-waves, and we can remain in the sunshine without suffering from the heat.

This is how the water-vapour shields us by day, but by night it is still more useful. During the day our earth and the air near it have been storing up the heat which has been poured down on them, and at night, when the sun goes down, all this heat begins to escape again. Now, if there were no vapour in the air, this heat would rush back into space so rapidly that the ground would become cold and frozen even on a summer's night, and all but the most hardy plants would die. But the vapour which formed a veil against the sun in the day, now forms a still more powerful veil against the escape of the heat by night. It shuts in the heat- waves, and only allows them to make their way slowly upwards from the earth - thus producing for us the soft, balmy nights of summer and preventing all life being destroyed in the winter.
Perhaps you would scarcely imagine at first that it is this screen of vapour which determines whether or not we shall have dew upon the ground. Have you ever thought why dew forms, or what power has been at work scattering the sparkling drops upon the grass? Picture to yourself that it has been a very hot summer's day, and the ground and the grass have been well warmed, and that the sun goes down in a clear sky without any clouds. At once the heat- waves which have been stored up in the ground, bound back into the air, and here some are greedily absorbed by the vapour, while others make their way slowly upwards. The grass, especially, gives out these heat-waves very quickly, because the blades, being very thin, are almost all surface. In consequence of this they part with their heat more quickly than they can draw it up from the ground, and become cold. Now the air lying just above the grass is full of invisible vapour, and the cold of the blades, as it touches them, chills the water- particles, and they are no longer able to hold apart, but are drawn together into drops on the surface of the leaves.
We can easily make artificial dew for ourselves. I have here a bottle of ice which has been kept outside the window. When I bring it into the warm room a mist forms rapidly outside the bottle. This mist is composed of water-drops, drawn out of the air of the room, because the cold glass chilled the air all round it, so that it gave up its invisible water to form dew-drops. Just in this same way the cold blades of grass chill the air lying above them, and steal its vapour.
But try the experiment, some night when a heavy dew is expected, of spreading a thin piece of muslin over some part of the grass, supporting it at the four corners with pieces of stick so that it forms an awning. Though there may be plenty of dew on the grass all round, yet under this awning you will find scarcely any. The reason of this is that the muslin checks the heat-waves as they rise from the grass, and so the grass-blades are not chilled enough to draw together the water-drops on their

surface. If you walk out early in the summer mornings and look at the fine cobwebs flung across the hedges, you will see plenty of drops on the cobwebs themselves sparkling like diamonds; but underneath on the leaves there will be none, for even the delicate cobweb has been strong enough to shut in the heat-waves and keep the leaves warm.

Again, if you walk off the grass on to the gravel path, you find no dew there. Why is this? Because the stones of the gravel can draw up heat from the earth below as fast as they give it out, and so they are never cold enough to chill the air which touches them. On a cloudy night also you will often find little or no dew even on the grass. The reason of this is that the clouds give back heat to the earth, and so the grass does not become chilled enough to draw the water-drops together on its surface. But after a hot, dry day, when the plants are thirsty and there is little hope of rain to refresh them, then they are able in the evening to draw the little drops from the air and drink them in before the rising sun comes again to carry them away.

But our rain-drop undergoes other changes more strange than these. Till now we have been imagining it to travel only where the temperature is moderate enough for it to remain in a liquid state as water. But suppose that when it is drawn up into the air it meets with such a cold blast as to bring it to the freezing point. If it falls into this blast when it is already a drop, then it will freeze into a hailstone, and often on a hot summer's day we may have a severe hailstorm, because the rain-drops have crossed a bitterly cold wind as they were falling, and have been frozen into round drops of ice.

But if the water-vapour reaches the freezing air while it is still an invisible gas, and before it has been drawn into a drop, then its history is very different. The ordinary force of cohesion has then no power over the particles to make them into watery globes, but its place is taken by the fairy process of "crystallization," and they are formed into beautiful white flakes, to fall in a snow-shower. I want you to picture this process to yourselves, for if once you can take an interest in the wonderful power of nature to build up crystals, you will be astonished how often you will meet with instances of it, and what pleasure it will add to your life.

The particles of nearly all substances, when left free and not hurried, can build themselves into crystal forms. If you melt salt in water and then let all the water evaporate slowly, you will get salt-crystals; — beautiful cubes of transparent salt all built on the same pattern. The same is true of sugar; and if you will look at the spikes of an ordinary stick of sugar-candy, such as I have here, you will see the kind of crystals which sugar forms. You may even pick out such shapes as these from the common crystallized brown sugar in the sugar basin, or see them with a

magnifying glass on a lump of white sugar.
But it is not only easily melted substances such as sugar and salt which form crystals. The beautiful stalactite grottos are all made of crystals of lime. Diamonds are crystals of carbon, made inside the earth. Rock-crystals, which you know probably under the name of Irish diamonds, are crystallized quartz; and so, with slightly different colourings, are agates, opals, jasper, onyx, cairngorms, and many other precious stones. Iron, copper, gold, and sulphur, when melted and cooled slowly build themselves into crystals, each of their own peculiar form, and we see that there is here a wonderful order, such as we should never have dreamt of, if we had not proved it. If you possess a microscope you may watch the growth of crystals yourself by melting some common powdered nitre in a little water till you find that no more will melt in it. Then put a few drops of this water on a warm glass slide and place it under the microscope. As the drops dry you will see the long transparent needles of nitre forming on the glass, and notice how regularly these crystals grow, not by taking food inside like living beings, but by adding particle to particle on the outside evenly and regularly.

Week 12
Can we form any idea why the crystals build themselves up so systematically? Dr. Tyndall says we can, and I hope by the help of these small bar magnets to show you how he explains it. These little pieces of steel, which I hope you can see lying on this white cardboard, have been rubbed along a magnet until they have become magnets themselves, and I can attract and lift up a needle with any one of them. But if I try to lift one bar with another, I can only do it by bringing certain ends together. I have tied a piece of red cotton (c, Fig. 21) round one end of each of the magnets, and if I bring two red ends together they will not cling together but roll apart. If, on the contrary, I put a red end against an end where there is not cotton, then the two bars cling together. This is because every magnet has two poles or points which are exactly opposite in character, and to distinguish them one is called the positive pole and the other the negative pole. Now when I bring two red ends, that is, two positive poles together, they drive each other away. See! the magnet I am not holding runs away from the other. But if I bring a red end and a black end, that is, a positive and a negative end together, then they are attracted and cling. I will make a triangle (A, Fig. 21) in which a black end and a red end always come together, and you see the triangle holds together. But now if I take off the lower bar and turn it (B, Fig. 21) so

that two red ends and two black ends come together, then this bar actually rolls back from the others down the cardboard. If I were to break these bars into a thousand pieces, each piece would still have two poles, and if they were scattered about near each other in such a way that they were quite free to move, they would arrange themselves always so two different poles came together.

Now picture to yourselves that all the particles of those substances which form crystals have poles like our magnets, then you can imagine that when the heat which held them apart is withdrawn and the particles come very near together, they will arrange themselves according to the attraction of their poles and so build up regular and beautiful patterns.

So, if we could travel up to the clouds where this fairy power of crystallization is at work, we should find the particles of water-vapour in a freezing atmosphere being built up into minute solid crystals of snow. If you go out after a snow-shower and search carefully, you will see that the snow-flakes are not mere lumps of frozen water, but beautiful six-pointed crystal stars, so white and pure that when we want to speak of anything being spotlessly white, you say that it is "white as snow." Some of these crystals are simply flat slabs with six sides, others are stars with six rods or spikes springing from the centre, others with six spikes each formed like a delicate fern. No less than a thousand different forms of delicate crystals have been found among snowflakes, but though there is such a great variety, yet they are all built on the six-sided and six-pointed plan, and are all rendered dazzlingly white by the reflection of the light from the faces of the crystals and the tiny air-bubbles built up within them. This, you see, is why, when the snow melts, you have only a little dirty water in your hand; the crystals are gone and there are no more air-bubbles held prisoners to act as looking-glasses to the light. Hoar-frost is also made up of tiny water-crystals, and is nothing more than frozen dew hanging on the blades of grass and from the trees.

But how about ice? Here, you will say, is frozen water, and yet we see no crystals, only a clear transparent mass. Here, again, Dr. Tyndall helps us. He says (and as I have proved it true, so may you for yourselves, if you will) that if you take a magnifying glass, and look down on the surface of ice on a sunny day, you will see a number of dark, six-sided stars, looking like flattened flowers, and in the centre of each a bright spot. These flowers, which are seen when the ice is melting, are our old friends the crystal stars turning into water, and the bright spot in the middle is a bubble of empty space, left because the watery flower does not fill up as much room as the ice of the crystal star did.

And this leads us to notice that ice always takes up more room than water, and that this is the reason why our water-pipes burst in severe frosts; for as the water freezes it expands with great force, and the pipe is cracked, and then when the thaw comes on , and the water melts again, it pours through the crack it has made.

It is not difficult to understand why ice should take more room; for we know that if we were to try to arrange bricks end to end in star-like shapes, we must leave some spaces between, and could not pack them so closely as if they lay side by side. And so, when this giant force of crystallization constrains the atoms of frozen water to grow into star-like forms, the solid mass must fill more room than the liquid water, and when the star melts, this space reveals itself to us in the bright spot of the centre.

We have now seen our drop of water under all its various forms of invisible gas, visible steam, cloud, dew, hoar-frost, snow, and ice, and we have only time shortly to see it on its travels, not merely up and down, as hitherto, but round the world.

We must first go to the sea as the distillery, or the place from which water is drawn up invisibly, in its purest state, into the air; and we must go chiefly to the seas of the tropics, because here the sun shines most directly all the year round, sending heat-waves to shake the water-particles asunder. It has been found by experiment that, in order to turn 1 lb. of water into vapour, as much heat must be used as is required to melt 5 lbs. of iron; and if you consider for a moment how difficult iron is to melt, and how we can keep an iron poker in a hot fire and yet it remains solid, this will help you to realize how much heat the sun must pour down in order to carry off such a constant supply of vapour from the tropical seas.

Now, when all this vapour is drawn up into the air, we know that some of it will form into clouds as it gets chilled high up in the sky, and then it will pour down again in those tremendous floods of rain which occur in the tropics.

But the sun and air will not let it all fall down at once, and the winds which are blowing from the equator to the poles carry large masses of it away with them. Then, as you know, it will depend on many things how far this vapour is carried. Some of it, chilled by cold blasts, or by striking on cold mountain tops, as it travels northwards, will fall in rain in Europe and Asia, while that which travels southwards may fall in South America, Australia, or New Zealand, or be carried over the sea to the South Pole. Wherever it falls on the land as rain, and is not used by plants, it will do one of two things; either it will run down in streams and form brooks and rivers, and so at last find its way back to the sea, or it will sink deep in the earth till it comes upon some hard rock through which it cannot get, and then, being hard pressed by the water coming on behind, it will rise up again through cracks, and come to the surface as a spring. These springs, again, feed rivers, sometimes above- ground, sometimes for long distances under-ground; but one way or another at last the whole drains back into the sea.

But if the vapour travels on till it reaches high mountains in cooler lands, such as the Alps of Switzerland; or is carried to the poles and to such countries as Greenland or the Antarctic Continent, then it will come down as snow, forming immense snow- fields. And here a curious change takes place in it. If you make an ordinary snowball and work it firmly together, it becomes very hard, and if you then press it forcibly into a mould you can turn it into transparent ice. And in the same way the snow which falls in Greenland and on the high mountains of Switzerland becomes very firmly pressed together, as it slides down into the valleys. It is like a crowd of people passing from a broad thoroughfare into a narrow street. As the valley grows narrower and narrower the great mass of snow in front cannot move down quickly, while more and more is piled up by the snowfall behind, and the crowd and crush grow denser and denser. In this way the snow is pressed together till the air that was hidden in its crystals, and which gave it its beautiful whiteness, is all pressed out, and the snow-crystals themselves are squeezed into one solid mass of pure, transparent ice.

Then we have what is called a "glacier," or river of ice, and this solid river comes creeping down till, in Greenland, it reaches the edge of the sea. There it is pushed over the brink of the land, and large pieces snap off, and we have "icebergs." These icebergs - made, remember, of the same water which was first draw up from the tropics - float on the wide sea, and melting in its warm currents, topple over and over* (A floating iceberg must have about eight times as much ice under the water as it has above, and therefore, when the lower part melts in a warm current, the iceberg loses its balance and tilts over, so as to rearrange itself round the centre of gravity.) till they disappear and mix with the water, to be carried back again to the warm ocean from which they first started. In Switzerland the glaciers cannot reach the sea, but they move down into the valleys till they come to a warmer region, and there the end of the glacier melts, and flows away in a stream. The Rhone and many other rivers are fed by the glaciers of the Alps; and as these rivers flow into the sea, our drop of water again finds its way back to its home.

But when it joins itself in this way to its companions, from whom it was parted for a time, does it come back clear and transparent as it left them? From the iceberg it does indeed return pure and clear; for the fairy Crystallization will have no impurities, not even salt, in her ice-crystals, and so as they melt they give back nothing but pure water to the sea. Yet even icebergs bring down earth and stones frozen into the bottom of the ice, and so they feed the sea with mud.

But the drops of water in rivers are by no means as pure as when they rose up into the sky. We shall see in the next lecture how rivers carry down not only sand and mud all along their course, but even solid matter such as salt, lime, iron, and flint, dissolved in the clear water, just as sugar is dissolved, without our being able to see it. The water, too, which has sunk down into the earth, takes up much matter as it travels along. You all know that the water you drink from a spring is very different from rain-water, and you will often find a hard crust at the bottom of kettles and in boilers, which is formed of the carbonate of lime which is driven out of the clear water when it is boiled. The water has become "hard" in consequence of having picked up and dissolved the carbonate of lime on its way through the earth, just in the same way as water would become sweet if you poured it through a sugar-cask. You will also have heard of iron-springs, sulphur-springs, and salt-springs, which come out of the earth, even if you have never tasted any of them, and the water of all these springs finds its way back at last to the sea.

And now, can you understand why sea-water should taste salt and bitter? Every drop of water which flows from the earth to the sea carries something with it. Generally, there is so little of any substance in the water that we cannot taste it, and we call it pure water; but the purest of spring or river-water has always some solid matter dissolved in it, and all this goes to the sea. Now, when the sun-waves come to take the water out of the sea again, they will have nothing but the pure water itself; and so all these salts and carbonates and other solid substances are left behind, and we taste them in sea-water.

Some day, when you are at the seaside, take some extra water and set it on the hob till a great deal has simmered gently away, and the liquid is very thick. Then take a drop of this liquid, and examine it under a microscope. As it dries up gradually, you will see a number of crystals forming, some square - and these will be crystals of ordinary salt; some oblong - these will be crystals of gypsum or alabaster; and others of various shapes. Then, when you see how much matter from the land is contained in sea-water, you will no longer wonder that the sea is salt; on the contrary, you will ask, Why does it not grow salter every year?

The answer to this scarcely belongs to our history of a drop of water, but I must just suggest it to you. In the sea are numbers of soft-bodied animals, like the jelly animals which form the coral, which require hard material for their shells or the solid branches on which they live, and they are greedily watching for these atoms of lime, of flint, or magnesia, and of other substances brought down into the sea. It is with lime and magnesia that the tiny chalk-builders form their beautiful shells, and the coral animals their skeletons, while another class of builders use the flint; and when these creatures die, their remains go to form fresh land at the bottom of the sea; and so, though the earth is being washed away by the rivers and springs it is being built up again, out of the same materials, in the depths of the great ocean.

And now we have reached the end of the travels of our drop of water. We have seen it drawn up by the fairy "heat," invisible into the sky; there fairy "cohesion" seized it and formed it into water-drops and the giant, "gravitation," pulled it down again to the earth. Or, if it rose to freezing regions, the fairy of "crystallization" built it up into snow-crystals, again to fall to the earth, and either to be melted back into water by heat, or to slide down the valleys by force of gravitation, till it became squeezed into ice. We have detected it, when invisible, forming a veil round our earth, and keeping off the intense heat of the sun's rays by day, or shutting it in by night. We have seen it chilled by the blades of grass, forming sparkling dew-drops or crystals of hoar-frost, glistening in the early morning sun; and we have seen it in the dark underground, being drunk up greedily by the roots of plants. We have started with it from the tropics, and travelled over land and sea, watching it forming rivers, or flowing underground in springs, or moving onwards to the high mountains or the poles, and coming back again in glaciers and icebergs. Through all this, while it is being carried hither and thither by invisible power, we find no trace of its becoming worn out, or likely to rest from its labours. Ever onwards it goes, up and down, and round and round the world, taking many forms, and performing many wonderful feats. We have seen some of the work that it does, in refreshing the air, feeding the plants, giving us clear, sparkling water to drink, and carrying matter to the sea; but besides this, it does a wonderful work in altering all the face of our earth. This work we shall consider in the next lecture, on "The two great Sculptors - Water and Ice."

Week 13

LECTURE V. THE TWO GREAT SCULPTORS - WATER AND ICE.

In our last lecture we saw that water can exist in three forms:— 1st, as an invisible vapour; 2nd, as liquid water; 3rd, as solid snow and ice.

To-day we are going to take the two last of these forms, water and ice, and speak of them as sculptors.

To understand why they deserve this name we must first consider what the work of a sculptor is. If you go into a statuary yard you will find there large blocks of granite, marble, and other kinds of stone, hewn roughly into different shapes; but if you pass into the studio, where the sculptor himself is at work you will find beautiful statues, more or less finished; and you will see that out of rough blocks of stone he has been able to cut images which look like living forms. You can even see by their faces whether they are intended to be sad, or thoughtful, or gay, and by their attitude whether they are writhing in pain, or dancing with joy, or resting peacefully. How has all this history been worked out from the shapeless stone? It has been done by the sculptor's chisel. A piece chipped off here, a wrinkle cut there, a smooth surface rounded off in another place, so as to give a gentle curve; all these touches gradually shape the figure and mould it out of the rough stone, first into a rude shape and afterwards, by delicate strokes, into the form of a living being.

Now, just in the same way as the wrinkles and curves of a statue are cut by the sculptor's chisel, so the hills and valleys, the steep slopes and gentle curves on the face of our earth, giving it all its beauty, and the varied landscapes we love so well, have been cut out by water and ice passing over them. It is true that some of the greater wrinkles of the earth, the lofty mountains, and the high masses of land which rise above the sea , have been caused by earthquakes and shrinking of the earth. We shall not speak of these to-day, but put them aside as belonging to the rough work of the statuary yard. But when once these large masses are put ready for water to work upon, then all the rest of the rugged wrinkles and gentle slopes which make the country so beautiful are due to water and ice, and for this reason I have called them "sculptors."

Go for a walk in the country, or notice the landscape as you travel on a railway journey. You pass by hills and through valleys, through narrow steep gorges cut in hard rock, or through wild ravines up the sides of which you can hardly scramble. Then you come to grassy slopes and to smooth plains across which you can look for miles without seeing a hill; or, when you arrive at the seashore, you clamber into caves and grottos, and along dark narrow passages leading from one bay to another. All these - hills, valleys, gorges, ravines, slopes, plains, caves, grottos, and rocky shores - have been cut out by the water. Day by day and year by year, while everything seems to us to remain the same, this industrious sculptor is chipping away, a few grains here, a corner there, a large mass in another place, till he gives to the country its own peculiar scenery, just as the human sculptor gives expression to his statue.

Our work to-day will consist in trying to form some idea of the way in which water thus carves out the surface of the earth, and we will begin by seeing how much can be done by our old friends the rain-drops before they become running streams.

Everyone must have noticed that whenever rain falls on soft ground it makes small round holes in which it collects, and then sinks into the ground, forcing its way between the grains of earth. But you would hardly think that the beautiful pillars in Fig. 24 have been made entirely in this way by rain beating upon and soaking into the ground.

Where these pillars stand there was once a solid mass of clay and stones, into which the rain-drops crept, loosening the earthly particles; and then when the sun dried the earth again cracks were formed, so that the next shower loosened it still more, and carried some of the mud down into the valley below. But here and there large stones were buried in the clay, and where this happened the rain could not penetrate, and the stones became the tops of tall pillars of clay, washed into shape by the rain beating on its sides, but escaping the general destruction of the rest of the mud. In this way the whole valley has been carved out into fine pillars, some still having capping-stones, while others have lost them, and these last will soon be washed away. We have no such valleys of earth-pillars here in England, but you may sometimes see tiny pillars under bridges where the drippings have washed away the earth between the pebbles, and such small examples which you can observe for yourselves are quite as instructive as more important ones.

Another way in which rain changes the surface of the earth is by sinking down through loose soil from the top of a cliff to a depth of many feet till it comes to solid rock, and then lying spread over a wide apace. Here it makes a kind of watery mud, which is a very unsafe foundation for the hill of earth above it, and so after a time the whole mass slips down and makes a fresh piece of land at the foot of the cliff. If you have ever been at the Isle of Wight you will have seen an undulating strip of ground, called the Undercliff, at Ventnor and other places, stretching all along the sea below the high cliffs. This land was once at the top of the cliff, and came down by succession of landslips such as we have been describing. A very great landslip of this kind happened in the memory of living people, at Lyme Regis, in Dorsetshire, in the year 1839.

You will easily see how in forming earth-pillars and causing landslips rain changes the face of the country, but these are only rare effects of water. It is when the rain collects in brooks and forms rivers that it is most busy in sculpturing the land. Look out some day into the road or the garden where the ground slopes a little, and watch what happens during a shower of rain. First the rain-drops run together in every little hollow of the ground, then the water begins to flow along any ruts or channels it can find, lying here and there in pools, but always making its way gradually down the slope. Meanwhile from other parts of the ground little rills are coming, and these all meet in some larger ruts where the ground is lowest, making one great stream, which at last empties itself into the gutter or an area, or finds its way down some grating.

Now just this, which we can watch whenever a heavy shower of rain comes down on the road, happens also all over the world. Up in the mountains, where there is always a great deal of rain, little rills gather and fall over the mountain sides, meeting in some stream below. Then, as this stream flows on, it is fed by many runnels of water, which come from all parts of the country, trickling along ruts, and flowing in small brooks and rivulets down the gentle slope of the land till they reach the big stream, which at last is important enough to be called a river. Sometimes this river comes to a large hollow in the land and there the water gathers and forms a lake; but still at the lower end of this lake out it comes again, forming a new river, and growing and growing by receiving fresh streams until at last it reaches the sea.

The River Thames, which you all know, and whose course you will find clearly described in Mr. Huxley's 'Physiography,' drains in this way no less than one-seventh of the whole of England. All the rain which falls in Berkshire, Oxfordshire, Middlesex, Hertfordshire, Surrey, the north of Wiltshire and north-west of Kent, the south of Buckinghamshire and of Gloucestershire, finds its way into the Thames; making an area of 6160 square miles over which every rivulet and brook trickle down to the one great river, which bears them to the ocean. And so with every other area of land in the world there is some one channel towards which the ground on all sides slopes gently down, and into this channel all the water will run, on its way to the sea.

But what has this to do with sculpture or cutting out of valleys? If you will only take a glass of water out of any river, and let it stand for some hours, you will soon answer this question for yourself. For you will find that even from river water which looks quite clear, a thin layer of mud will fall to the bottom of the glass, and if you take the water when the river is swollen and muddy you will get quite a thick deposit. This shows that the brooks, the streams, and the rivers wash away the land as they flow over it and carry it from the mountains down to the valleys, and from the valleys away out into the sea.

But besides earthly matter, which we can see, there is much matter dissolved in the water of rivers (as we mentioned in the last lecture), and this we cannot see.

If you use water which comes out of a chalk country you will find that after a time the kettle in which you have been in the habit of boiling this water has a hard crust on its bottom and sides, and this crust is made of chalk or carbonate of lime, which the water took out of the rocks when it was passing through them. Professor Bischoff has calculated that the river Rhine carries past Bonn every year enough carbonate of lime dissolved in its water to make 332,000 million oyster-shells, and that if all these shells were built into a cube it would measure 560 feet.

Week 14

Imagine to yourselves the whole of St. Paul's churchyard filled with oyster-shells, built up in a large square till they reached half as high again as the top of the cathedral, then you will have some idea of the amount of chalk carried invisibly past Bonn in the water of the Rhine every year.

Since all this matter, whether brought down as mud or dissolved, comes from one part of the land to be carried elsewhere or out to sea, it is clear that some gaps and hollows must be left in the places from which it is taken. Let us see how these gaps are made. Have you ever clambered up the mountainside, or even up one of those small ravines in the hillside, which have generally a little stream trickling through them? If so, you must have noticed the number of pebbles, large and small, lying in patches here and there in the stream, and many pieces of broken rock, which are often scattered along the sides of the ravine; and how, as you climb, the path grows steeper, and the rocks become rugged and stick out in strange shapes.

The history of this ravine will tell us a great deal about the carving of water. Once it was nothing more than a little furrow in the hillside down which the rain found its way in a thin thread-like stream. But by and by, as the stream carried down some of the earth, and the furrow grew deeper and wider, the sides began to crumble when the sun dried up the rain which had soaked in. Then in winter, when the sides of the hill were moist with the autumn rains, frost came and turned the water to ice, and so made the cracks still larger, and the swollen steam rushing down, caught the loose pieces of rock and washed them down into its bed. Here they were rolled over and over, and grated against each other, and were ground away till they became rounded pebbles, such as lie in the foreground of the picture (Fig. 25); while the grit which was rubbed off them was carried farther down by the stream. And so in time this became a little valley, and as the stream cut it deeper and deeper, there was room to clamber along the sides of it, and ferns and mosses began to cover the naked stone, and small trees rooted themselves along the banks, and this beautiful little nook sprang up on the hill-side entirely by the sculpturing of water.

Shall you not feel a fresh interest in all the little valleys, ravines, and gorges you meet with in the country, if you can picture them being formed in this way year by year? There are many curious differences in them which you can study for yourselves. Some will be smooth, broad valleys and here the rocks have been soft and easily worn, and water trickling down the sides of the first valley has cut other channels so as to make smaller valleys running across it. In other places there will be narrow ravines, and here the rocks have been hard, so that they did not wear away gradually, but broke off and fell in blocks, leaving high cliffs on each side. In some places you will come to a beautiful waterfall, where the water has tumbled over a steep cliff, and then eaten its way back, just like a saw cutting through a piece of wood.

There are two things in particular to notice in a waterfall like this. First, how the water and spray dash against the bottom of the cliff down which it falls, and grind the small pebbles against the rock. In this way the bottom of the cliff is undermined, and so great pieces tumble down from time to time, and keep the fall upright instead of its being sloped away at the top, and becoming a mere steam. Secondly, you may often see curious cup-shaped holes, called "pot-holes," in the rocks on the sides of a waterfall, and these also are concerned in its formation. In these holes you will generally find two or three small pebbles, and you have here a beautiful example of how water uses stones to grind away the face of the earth. These holes are made entirely by the falling water eddying round and round in a small hollow of the rock, and grinding the pebbles which it has brought down, against the bottom and sides of this hollow, just as you grind round a pestle in a mortar. By degrees the hole grows deeper and deeper and though the first pebbles are probably ground down to powder, others fall in, and so in time there is a great hole perforated right through, helping to make the rock break and fall away.

In this and other ways the water works its way back in a surprising manner. The Isle of Wight gives us some good instances of this; Alum Bay Chine and the celebrated Blackgang Chine have been entirely cut out by waterfalls. But the best know and most remarkable example is the Niagara Falls, in America. Here, the River Niagara first wanders through a flat country, and then reaches the great Lake Erie in a hollow of the plain. After that, it flows gently down for about fifteen miles, and then the slope becomes greater and it rushes on to the Falls of Niagara. These falls are not nearly so high as many people imagine, being only 165 feet, or about half the height of St. Paul's Cathedral, but they are 2700 feet or nearly half-a-mile wide, and no less than 670,000 tons of water fall over them every minute, making magnificent clouds of spray.

Sir Charles Lyell, when he was at Niagara, came to the conclusion that, taking one year with another, these falls eat back the cliff at the rate of about one foot a year, as you can easily imagine they would do, when you think with what force the water must dash against the bottom of the falls. In this way a deep cleft has been cut right back from Queenstown for a distance of seven miles, to the place where the falls are now. This helps us a little to understand how very slowly and gradually water cuts its way; for if a foot a year is about the average of the waste of the rock, it will have taken more than thirty-five thousand years for that channel of seven miles to be made.

But even this chasm cut by the falls of Niagara is nothing compared with the canyons of Colarado. Canyon is a Spanish word for a rocky gorge, and these gorges are indeed so grand, that if we had not seen in other places what water can do, we should never have been able to believe that it could have cut out these gigantic chasms. For more than three hundred miles the River Colorado, coming down from the Rocky Mountains, has eaten its way through a country made of granite and hard beds of limestone and sandstone, and it has cut down straight through these rocks, leaving walls from half-a-mile to a mile high, standing straight up from it. The cliffs of the Great Canyon, as it is called, stretch up for more than a mile above the river which flows in the gorge below! Fancy yourselves for a moment in a boat on this river, as shown in Figure 27, and looking up at these gigantic walls of rock towering above you. Even half-way up them, a man, if he could get there, would be so small you could not see him without a telescope; while the opening at the top between the two walls would seem so narrow at such an immense distance that the sky above would have the appearance of nothing more than a narrow streak of blue. Yet these huge chasms have not been made by any violent breaking apart of the rocks or convulsion of an earthquake. No, they have been gradually, silently, and steadily cut through by the river which now glides quietly in the wider chasms, or rushes rapidly through the narrow gorges at their feet.

"No description," says Lieutenant Ives, one of the first explorers of this river, "can convey the idea of the varied and majestic grandeur of this peerless waterway. Wherever the river turns, the entire panorama changes. Stately facades, august cathedrals, amphitheatres, rotundas, castellated walls, and rows of time-stained ruins, surmounted by every form of tower, minaret, dome and spire, have been moulded from the cyclopean masses of rock that form the mighty defile." Who will say, after this, that water is not the grandest of all sculptors, as it cuts through hundreds of miles of rock, forming such magnificent granite groups, not only unsurpassed but unequalled by any of the works of man?

But we must not look upon water only as a cutting instrument, for it does more than merely carve out land in one place, it also carries it away and lays it down elsewhere; and in this it is more like a modeller in clay, who smooths off the material from one part of his figure to put it upon another.

Running water is not only always carrying away mud, but at the same time laying it down here and there wherever it flows. When a torrent brings down stones and gravel from the mountains, it will depend on the size and weight of the pieces how long they will be in falling through the water. If you take a handful of gravel and throw it into a glass full of water, you will notice that the stones in it will fall to the bottom at once, the grit and coarse sand will take longer in sinking, and lastly, the fine sand will be an hour or two in settling down, so that the water becomes clear. Now, suppose that this gravel were sinking in the water of a river. The stones would be buoyed up as long as the river was very full and flowed very quickly, but they would drop through sooner than the coarse sand. The coarse sand in its turn would begin to sink as the river flowed more slowly, and would reach the bottom while the fine sand was still borne on. Lastly, the fine sand would sink through very, very slowly, and only settle in comparatively still water.

From this it will happen that stones will generally lie near to the bottom of torrents at the foot of the banks from which they fall, while the gravel will be carried on by the stream after it leaves the mountains. This too, however, will be laid down when the river comes into a more level country and runs more slowly. Or it may be left together with the finer mud in a lake, as in the lake of Geneva, into which the Rhone flows laden with mud and comes out at the other end clear and pure. But if no lake lies in the way the finer earth will still travel on, and the river will take up more and more as it flows, till at last it will leave this too on the plains across which it moves sluggishly along, or will deposit it at its mouth when it joins the sea.

Week 15

You all know the history of the Nile; how, when the rains fall very heavily in March and April in the mountains of Abyssinia, the river comes rushing down and brings with it a load of mud which it spreads out over the Nile valley in Egypt. This annual layer of mud is so thin that it takes a thousand years for it to become 2 or 3 feet thick; but besides that which falls in the valley a great deal is taken to the mouth of the river and there forms new land, making what is called the "Delta" of the Nile. Alexandria, Rosetta, and Damietta, are towns which are all built on land made of Nile mud which was carried down ages and ages ago, and which has now become firm and hard like the rest of the country. You will easily remember other deltas mentioned in books, and all these are made of the mud carried down from the land to the sea. The delta of the Ganges and Brahmapootra in India, is actually as large as the whole of England and Wales, (58,311 square miles.) and the River Mississippi in America drains such a large tract of country that its delta grows, Mr. Geikie tells us, at the rate of 86 yards in year.

All this new land laid down in Egypt, in India, in America, and in other places, is the work of water. Even on the Thames you may see mud-banks, as at Gravesend, which are made of earth brought from the interior of England. But at the mouth of the Thames the sea washes up very strongly every tide, and so it carries most of the mud away and prevents a delta growing up there. If you will look about when you are at the seaside, and notice wherever a stream flows down into the sea, you may even see little miniature deltas being formed there, though the sea generally washes them away again in a few hours, unless the place is well sheltered.

This, then, is what becomes of the earth carried down by rivers. Either on plains, or in lakes, or in the sea, it falls down to form new land. But what becomes of the dissolved chalk and other substances? We have seen that a great deal of it is used by river and sea animals to build their shells and skeletons, and some of it is left on the surface of the ground by springs when the water evaporates. It is this carbonate of lime which forms a hard crust over anything upon which it may happen to be deposited, and then these things are called "petrified."

But it is in the caves and hollows of the earth that this dissolved matter is built up into the most beautiful forms. If you have ever been to Buxton in Derbyshire, you will probably have visited a cavern called Poole's Cavern, not far from there, which when you enter it looks as if it were built up entirely of rods of beautiful transparent white glass, hanging from the ceiling, from the walls, or rising up from the floor. In this cavern, and many others like it,*(See the picture at the head of the lecture.) water comes dripping through the roof, and as it falls carbonate of lime forms itself into a thin, white film on the roof, often making a complete circle, and then, as the water drips from it day by day, it goes on growing and growing till it forms a long needle-shaped or tube-shaped rod, hanging like an icicle. These rods are called stalactites, and they are so beautiful, as their minute crystals glisten when a light is taken into the cavern, that one of them near Tenby is called the "Fairy Chamber." Meanwhile, the water which drips on to the floor also leaves some carbonate of lime where it falls, and this forms a pillar, growing up towards the roof, and often the hanging stalactites and the rising pillars (called stalagmites) meet in the middle and form one column. And thus we see that underground, as well as aboveground, water moulds beautiful forms in the crust of the earth. At Adelsberg, near Trieste, there is a magnificent stalactite grotto made of a number of chambers one following another, with a river flowing through them; and the famous Mammoth Cave of Kentucky, more than ten miles long, is another example of these wonderful limestone caverns.

But we have not yet spoken of the sea, and this surely is not idle in altering the shape of the land. Even the waves themselves in a storm wash against the cliffs and bring down stones and pieces of rock on to the shore below. And they help to make cracks and holes in the cliffs, for as they dash with force against them they compress the air which lies in the joints of the stone and cause it to force the rock apart, and so larger cracks are made and the cliff is ready to crumble.

It is, however, the stones and sand and pieces of rock lying at the foot of the cliff which are most active in wearing it away. Have you never watched the waves breaking upon a beach in a heavy storm? How they catch up the stones and hurl them down again, grinding them against each other! At high tide in such a storm these stones are thrown against the foot of the cliff, and each blow does something towards knocking away part of the rock, till at last, after many storms, the cliff is undermined and large pieces fall down. These pieces are in their turn ground down to pebbles which serve to batter against the remaining rock. Professor Geikie tells us that the waves beat in a storm against the Bell Rock Lighthouse with as much force as if you dashed a weight of 3 tons against every square inch of the rock, and Stevenson found stones of 2 tons' weight which had been thrown during storms right over the ledge of the lighthouse. Think what force there must be in waves which can lift up such a rock and throw it, and such force as this beats upon our sea-coasts and eats away the land.

Fig. 28 is a sketch on the shores of Arbroath which I made some years ago. You will not find it difficult to picture to yourselves how the sea has eaten away these cliffs till some of the strongest pieces which have resisted the waves stand out by themselves in the sea. That cave in the left-hand corner ends in a narrow dark passage from which you come out on the other side of the rocks into another bay. Such caves as these are made chiefly by the force of the waves and the air, bringing down pieces of rock from under the cliff and so making a cavity, and then as the waves roll these pieces over and over and grind them against the sides, the hole is made larger. There are many places on the English coast where large pieces of the road are destroyed by the crumbling down of cliffs when they have been undermined by caverns such as these.

Thus, you see, the whole of the beautiful scenery of the sea - the shores, the steep cliffs, the quiet bays, the creeks and caverns - are all the work of the "sculptor" water; and he works best where the rocks are hardest, for there they offer him a good stout wall to batter, whereas in places where the ground is soft it washes down into a gradual gentle slope, and so the waves come flowing smoothly in and have no power to eat away the shore.

And now, what has Ice got to do with the sculpturing of the land? First, we must remember how much the frost does in breaking up the ground. The farmers know this, and always plough after a frost, because the moisture, freezing in the ground, has broken up the clods, and done half their work for them.

But this is not the chief work of ice. You will remember how we learnt in our last lecture that snow, when it falls on the mountains, gradually slides down into the valleys, and is pressed together by the gathering snow behind until it becomes moulded into a solid river of ice (see Fig. 29, Frontispiece). In Greenland and in Norway there are enormous ice-rivers or glaciers, and even in Switzerland some of them are very large. The Aletsch glacier, in the Alps, is fifteen miles long, and some are even longer than this. They move very slowly - on an average about 20 to 27 inches in the centre, and 13 to 19 inches at the sides every twenty-four hours, in the summer and autumn. How they move, we cannot stop to discuss now; but if you will take a slab of thin ice and rest it upon its two ends only, you can prove to yourself that ice does bend, for in a few hours you will find that its own weight has drawn it down in the centre, so as to form a curve. This will help you to picture to yourselves how glaciers can adapt themselves to the windings of the valley, creeping slowly onwards until they come down to a point where the air is warm enough to melt them, and then the ice flows away in a stream of water. It is very curious to see the number of little rills running down the great masses of ice at the glacier's mouth, bringing down with them gravel, and every now and then a large stone, which falls splashing into the stream below. If you look at the glacier in the Frontispiece, you will see that these stones come from those long lines of stones and boulders stretching along the sides and centre of the glacier. It is easy to understand where the stones at the side come from; for we have seen that damp and frost cause pieces to break off the surface of the rocks, and it is natural that these pieces should roll down the steep sides of the mountains on to the glacier. But the middle row requires some explanation. Look to the back of the picture, and you will see that this line of stones is made of two side rows, which come from the valleys above. Two glaciers, you see, have there joined into one, and so made a heap of stones all along their line of junction.

These stones are being continually, though slowly, conveyed by the glacier, from all the mountains along its sides, down to the place where it melts. Here it lets them fall, and they are gradually piled up till they form great walls of stone, which are called moraines. Some of the moraines left by the larger glaciers of olden time, in the country near Turin, form high hills, rising up even to 1500 feet.

Therefore, if ice did no more than carry these stone blocks, it would alter the face of the country; but it does much more than this. As the glacier moves along, it often cracks for a considerable way across its surface, and this crack widens and widens, until at last it becomes a great gaping chasm, or crevasse as it is called, so that you can look down it right to the bottom of the glacier. Into these crevasses large blocks of rock fall, and when the chasm is closed again as the ice presses on, these masses are frozen firmly into the bottom of the glacier, much in the same way as a steel cutter is fixed in the bottom of a plane. And they do just the same kind of work; for as the glacier slides down the valley, they scratch and grind the rocks underneath them, rubbing themselves away, it is true, but also scraping away the ground over which they move. In this way the glacier becomes a cutting instrument, and carves out the valleys deeper and deeper as it passes through them.

You may always know where a glacier has been, even if no trace of ice remains; for you will see rocks with scratches along them which have been cut by these stones; and even where the rocks have not been ground away, you will find them rounded like those in the left-hand of the Frontispiece, showing that the glacier- plane has been over them. These rounded rocks are called "roches moutonnees," because at the distance they look like sheep lying down.

You have only to look at the stream flowing from the mouth of a glacier to see what a quantity of soil it has ground off from the bottom of the valley; for the water is thick, and coloured a deep yellow by the mud it carries. This mud soon reaches the rivers into which the streams run; and such rivers as the Rhone and the Rhine are thick with matter brought down from the Alps. The Rhone leaves this mud in the Lake of Geneva, flowing out at the other end quite clear and pure. A mile and a half of land has been formed at the head of the lake since the time of the Romans by the mud thus brought down from the mountains.

Thus we see that ice, like water, is always busy carving out the surface of the earth, and sending down material to make new land elsewhere. We know that in past ages the glaciers were much larger than they are in our time; for we find traces of them over large parts of Switzerland where glaciers do not now exist, and huge blocks which could only have been carried by ice, and which are called "erratic blocks," some of them as big as cottages, have been left scattered over all the northern part of Europe. These blocks were a great puzzle to scientific men till, in 1840, Professor Agassiz showed that they must have been brought by ice all the way from Norway and Russia.

In those ancient days, there were even glaciers in England; for in Cumberland and in Wales you may see their work, in scratched and rounded rocks, and the moraines they have left. Llanberis Pass, so famous for its beauty, is covered with ice-scratches, and blocks are scattered all over the sides of the valley. There is one block high up on the right-hand slope of the valley, as you enter from the Beddgelert side, which is exactly poised upon another block, so that it rocks to and fro. It must have been left thus balanced when the ice melted round it. You may easily see that these blocks were carried by ice, and not by water, because their edges are sharp, whereas if they had been rolled in water, they would have been smoothed down.

We cannot here go into the history of that great Glacial Period long ago, when large fields of ice covered all the north of England; but when you read it for yourselves and understand the changes on the earth's surface which we can see being made by ice now, then such grand scenery as the rugged valleys of Wales, with large angular stone blocks scattered over them, will tell you a wonderful story of the ice of bygone times.

And now we have touched lightly on the chief ways in which water and ice carve out the surface of the earth. We have seen that rain, rivers, springs, the waves of the sea, frost, and glaciers all do their part in chiselling out ravines and valleys, and in producing rugged peaks or undulating plains - here cutting through rocks so as to form precipitous cliffs, there laying down new land to add to the flat country - in one place grinding stones to powder, in others piling them up in gigantic ridges. We cannot go a step into the country without seeing the work of water around us; every little gully and ravine tells us that the sculpture is going on; every stream, with its burden of visible or invisible matter, reminds us that some earth is being taken away and carried to a new spot. In our little lives we see indeed but the very small changes, but by these we learn how greater ones have been brought about, and how we owe the outline of all our beautiful scenery, with its hills and valleys, its mountains and plains, its cliffs and caverns, its quiet nooks and its grand rugged precipices, to the work of the "Two great sculptors, Water and

Ice."

Week 16
Lecture VI

THE VOICES OF NATURE AND HOW WE HEAR THEM

We have reached to-day the middle point of our course, and here we will make a new start. All the wonderful histories which we have been studying in the last five lectures have had little or nothing to do with living creatures. The sunbeams would strike on our earth, the air would move restlessly to and fro, the water-drops would rise and fall, the valleys and ravines would still be cut out by rivers , if there were no such thing as life upon the earth. But without living things there could be none of the beauty which these changes bring about. Without plants, the sunbeams, the air and the water would be quite unable to clothe the bare rocks, and without animals and man they could not produce light, or sound, or feeling of any kind.

In the next five lectures, however, we are going to learn something of the use living creatures make of the earth; and to- day we will begin by studying one of the ways in which we are affected by the changes of nature, and hear her voice.

We are all so accustomed to trust to our sight to guide us in most of our actions, and to think of things as we see them, that we often forget how very much we owe to sound. And yet Nature speaks to us so much by her gentle, her touching, or her awful sounds, that the life of a deaf person is even more hard to bear than that of a blind one.

Have you ever amused yourself with trying how many different sounds you can distinguish if you listen at an open window in a busy street? You will probably be able to recognize easily the jolting of the heavy wagon or dray, the rumble of the omnibus, the smooth roll of the private carriage and the rattle of the light butcher's cart; and even while you are listening for these, the crack of the carter's whip, the cry of the costermonger at his stall, and the voices of the passers-by will strike upon you ear. Then if you give still more close attention you will hear the doors open and shut along the street, the footsteps of the passengers, the scraping of the shovel of the mud-carts; nay, if he happen to stand near, you may even hear the jingling of the shoeblack's pence as he plays pitch and toss upon the pavement. If you think for a moment, does it not seem wonderful that you should hear all these sounds so that you can recognize each one distinctly while all the rest are going on around you? But suppose you go into the quiet country. Surely there will be silence there. Try some day and prove it for yourself, lie down on the grass in a sheltered nook and listen attentively. If there be ever so little wind stirring you will hear it rustling gently through the trees; or even if there is not this, it will be strange if you do not hear some wandering gnat buzzing, or some busy bee humming as it moves from flower to flower. Then a grasshopper will set up a chirp within a few yards of you, or, if all living creatures are silent, a brook not far off may be flowing along with a rippling musical sound. These and a hundred other noises you will hear in the most quiet country spot; the lowing of the cattle, the song of the birds, the squeak of the field-mouse, the croak of the frog, mingling with the sound of the woodman's axe in the distance, or the dash of some river torrent. And beside these quiet sounds, there are still other occasional voices of nature which speak to us from time to time. The howling of the tempestuous wind, the roar of the sea-waves in a storm, the crash of thunder, and the mighty noise of the falling avalanche; such sounds as these tell us how great and terrible nature can be.

Now, has it ever occurred to you to think what sounds is, and how it is that we hear all these things? Strange as it may seem, if there were no creature that could hear upon the earth, there would be no such thing as sound, though all these movements in nature were going on just as they are now.

Try and grasp this thoroughly, for it is difficult at first to make people believe it. Suppose you were stone-deaf, there would be no such thing as sound to you. A heavy hammer falling on an anvil would indeed shake the air violently, but since this air when it reached your ear would find a useless instrument, it could not play upon it. and it is this play on the drum of your ear and the nerves within it speaking to your brain which make sound. Therefore, if all creatures on or around the earth were without ears or nerves of hearing, there would be no instrument on which to play, and consequently there would be no such thing as sound. This proves that two things are needed in order that we may hear. First, the outside movement which plays on our hearing instrument; and, secondly, the hearing instrument itself.

First, then, let us try to understand what happens outside our ears. Take a poker and tie a piece of string to it, and holding the ends of the string to your ears, strike the poker against the fender. You will hear a very loud sound, for the blow will set all the particles of the poker quivering, and this movement will pass right along the string to the drum of your ear and play upon it.

Now take the string away from you ears, and hold it with your teeth. Stop your ears tight, and strike the poker once more against the fender. You will hear the sound quite as loudly and clearly as you did before, but this time the drum of your ear has not been agitated. How, then, has the sound been produced? In this case, the quivering movement has passed through your teeth into the bones of your hear, and from them into the nerves, and so produced sound in your brain. And now, as a final experiment, fasten the string to the mantelpiece, and hit it again against the fender. How much feebler the sound is this time, and how much sooner it stops! Yet still it reaches you, for the movement has come this time across the air to the drums of your ear.

Here we are back again in the land of invisible workers! We have all been listening and hearing ever since we were babies, but have we ever made any picture to ourselves of how sound comes to us right across a room or a field, when we stand at one end and the person who calls is at the other?

Since we have studied the "aerial ocean," we know that the air filling the space between us, though invisible, is something very real, and now all we have to do is to understand exactly how the movement crosses this air.

This we shall do most readily by means of an experiment made by Dr. Tyndall in his lectures on Sound. I have here a number of boxwood balls resting in a wooden tray which has a bell hung at the end of it. I am going to take the end ball and roll it sharply against the rest, and then I want you to notice carefully what happens. See! the ball at the other end has flow off and hit the bell, so that you hear it ring. Yet the other balls remain where they were before. Why is this? It is because each of the balls, as it was knocked forwards, had one in front of it to stop it and make it bound back again, but the last one was free to move on. When I threw this ball from my hand against the others, the one in front of it moved, and hitting the third ball, bounded back again; the third did the same to the fourth, the fourth to the fifth, and so on to the end of the line. Each ball thus came back to its place, but it passed the shock on to the last ball, and the ball to the bell. If I now put the balls close up to the bell, and repeat the experiment, you still hear the sound, for the last ball shakes the bell as if it were a ball in front of it.

Now imagine these balls to be atoms of air, and the bell your ear. If I clap my hands and so hit the air in front of them, each air-atom hits the next just as the balls did, and though it comes back to its place, it passes the shock on along the whole line to the atom touching the drum of your ear, and so you receive a blow. But a curious thing happens in the air which you cannot notice in the balls. You must remember that air is elastic, just as if there were springs between the atoms as in the diagram, Fig. 31, and so when any shock knocks the atoms forward, several of them can be crowded together before they push on those in front. Then, as soon as they have passed the shock on, they rebound and begin to separate again, and so swing to and fro till they come to rest. meanwhile the second set will go through just the same movements, and will spring apart as soon as they have passed the shock on to a third set, and so you will have one set of crowded atoms and one set of separated atoms alternately all along the line, and the same set will never be crowded two instants together.

You may see an excellent example of this in a luggage train in a railway station, when the trucks are left to bump each other till they stop. You will see three or four trucks knock together, then they will pass the shock on to the four in front, while they themselves bound back and separate as far as their chains will let them: the next four trucks will do the same, and so a kind of wave of crowded trucks passes on to the end of the train, and they bump to and fro till the whole comes to a standstill. Try to imagine a movement like this going on in the line of air- atoms, the drum of your ear being at the end. Those which are crowded together at that end will hit on the drum of your ear and drive the membrane which covers it inwards; then instantly the wave will change, these atoms will bound back, and the membrane will recover itself again, but only to receive a second blow as the atoms are driven forwards again, and so the membrane will be driven in and out till the air has settled down.

This you see is quite different to the waves of light which moves in crests and hollows. Indeed, it is not what we usually understand by a wave at all, but a set of crowdings and partings of atoms of air which follow each other rapidly across the air. A crowding of atoms is called a condensation, and a parting is called a rarefaction, and when we speak of the length of a wave of sound, we mean the distance between two condensations, or between two rarefactions.

Although each atom of air moves a very little way forwards and then back, yet, as a long row of atoms may be crowded together before they begin to part, a wave is often very long. When a man talks in an ordinary bass voice, he makes sound-waves from 8 to 12 feet long; a woman's voice makes shorter waves, from 2 to 4 feet long, and consequently the tone is higher, as we shall presently explain.

And now I hope that some one is anxious to ask why, when I clap my hands, anyone behind me or at the side, can hear it as well or nearly as well as you who are in front. This is because I give a shock to the air all round my hands, and waves go out on all sides, making as it were gloves of crowdings and partings widening and widening away from the clap as circles widen on a pond. Thus the waves travel behind me, above me, and on all sides, until they hit the walls, the ceiling, and the floor of the room, and wherever you happen to be, they hit upon your ear.

Week 17

If you can picture to yourself these waves spreading out in all directions, you will easily see why sound grows fainter at the distance. Just close round my hands when I clap them, there is a small quantity of air, and so the shock I give it is very violent, but as the sound-waves spread on all sides they have more and more air to move, and so the air-atoms are shaken less violently and strike with less force on your ear.

If we can prevent the sound-wave from spreading, then the sound is not weakened. The Frenchman Biot found that a low whisper could be heard distinctly for a distance of half a mile through a tube, because the waves could not spread beyond the small column of air. But unless you speak into a small space of some kind, you cannot prevent the waves going out from you in all directions.

Try and imagine that you see these waves spreading all round me now and hitting on your ears as they pass, then on the ears of those behind you, and on and on in widening globes till they reach the wall. What will happen when they get there? If the wall were thin, as a wooden partition is, they would shake it, and it again would shake the air on the other side, and so anyone in the next room would have the sound of my voice brought to their ear.

But something more will happen. In any case the sound-waves hitting against the wall will bound back from it just as a ball bounds back when thrown against anything, and so another set of sound-waves reflected from the wall will come back across the room. If these waves come to your ear so quickly that they mix with direct waves, they help to make the sound louder in this room than you would in the open air, for the "Ha" from my mouth and a second "Ha" from the wall come to your ear so instantaneously that they make one sound. This is why you can often hear better at the far end of a church when you stand against a screen or a wall, then when you are half-way up the building nearer to the speaker, because near the wall the reflected waves strike strongly on your ear and make the sound louder.

Sometimes, when the sound comes from a great explosion, these reflected waves are so strong that they are able to break glass. In the explosion of gunpowder in St. John's Wood, many houses in the back streets had their windows broken; for the sound-waves bounded off at

angles from the walls and struck back upon them.
Now suppose the wall were so far behind you that the reflected sound-waves only hit upon your ear after those coming straight from me had died away; then you would hear the sound twice, "Ha" from me and "Ha" from the wall, and here you have an echo, "Ha, ha." In order for this to happen in ordinary air, you must be standing at least 56 feet away from the point from which the waves are reflected, for then the second blow will come one-tenth of a second after the first one, and that is long enough for you to feel them separately.* Miss C. A. Martineau tells a story of a dog which was terribly frightened by an echo. Thinking another dog was barking, he ran forward to meet him, and was very much astonished, when, as he came nearer the wall, the echo ceased. I myself once knew a case of this kind, and my dog, when he could find no enemy, ran back barking, till he was a certain distance off, and then the echo of course began again. He grew so furious at last that we had great difficulty in preventing him from flying at a strange man who happened to be passing at the time. (*Sound travels 1120 feet in a second, in air of ordinary temperature, and therefore 112 feet in the tenth of a second. Therefore the journey of 56 feet beyond you to reach the wall and 56 feet to return, will occupy the sound-wave one-tenth of a second and separate the two sounds.)
Sometimes, in the mountains, walls of rock rise at some distance one behind another, and then each one will send back its echo a little later than the rock before it, so that the "Ha" which you give will come back as a peal of laughter. There is an echo in Woodstock Park which repeats the word twenty times. Again sometimes, as in the Alps, the sound-waves coming back rebound from mountain to mountain and are driven backwards and forwards, becoming fainter and fainter till they die away; these echoes are very beautiful.
If you are now able to picture to yourselves one set of waves going to the wall, and another set returning and crossing them, you will be ready to understand something of that very difficult question, How is it that we can hear many different sounds at one time and tell them apart?
Have you ever watched the sea when its surface is much ruffled, and noticed how, besides the big waves of the tide, there are numberless smaller ripples made by the wind blowing the surface of the water, or the oars of a boat dipping in it, or even rain- drops falling? If you have done this you will have seen that all these waves and ripples cross each other, and you can follow any one ripple with you eye as it goes on its way undisturbed by the rest. Or you may make beautiful crossing and recrossing ripples on a pond by throwing in two stones at a little distance from each other, and here too you can follow any one wave on to the

edge of the pond.

Now just in this way the waves of sound, in their manner of moving, cross and recross each other. You will remember too, that different sounds make waves of different lengths, just as the tide makes a long wave and the rain-drops tiny ones. Therefore each sound falls with its own peculiar wave upon your ear, and you can listen to that particular wave just as you look at one particular ripple, and then the sound becomes clear to you.

All this is what is going on outside your ear, but what is happening in your ear itself? How do these blows of the air speak to your brain? By means of the following diagram, Fig. 33, we will try to understand roughly our beautiful hearing instrument, the ear.

First, I want you to notice how beautifully the outside shell, or concha as it is called, is curbed round so that any movement of the air coming to it from the front is caught in it and reflected into the hole of the ear. Put your finger round your ear and feel how the gristly part is curved towards the front of your head. This concha makes a curve much like the curve a deaf man makes with his hand behind his ear to catch the sound. Animals often have to raise their ears to catch the sound well, but ours stand always ready. When the air-waves have passed in at the hole of your ear, they move all the air in the passage, which is called the auditory, or hearing, canal. This canal is lined with little hairs to keep out insects and dust, and the wax which collects in it serves the same purpose. But is too much wax collects, it prevents the air from playing well upon the drum, and therefore makes you deaf. Across the end of this canal, a membrane or skin called the tympanum is stretched, like the parchment over the head of a drum, and it is this membrane which moves to and fro as the air-waves strike on it. A violent box on the ear will sometimes break this delicate membrane, or injure it, and therefore it is very wrong to hit a person violently on the ear.

On the other side of this membrane, inside the ear, there is air, which fills the whole of the inner chamber and the tube, which runs down into the throat behind the nose, and is called the Eustachian tube after the man who discovered it. This tube is closed at the end by a valve which opens and shuts. If you breathe out strongly, and then shut your mouth and swallow, you will hear a little "click" in your ear. This is because in swallowing you draw the air out of the Eustachian tube and so draw in the membrane, which clicks as it goes back again. But unless you do this the tube and the whole chamber cavity behind the membrane remains full of air.

Now, as this membrane is driven to and fro by the sound-waves, it naturally shakes the air in the cavity behind it, and it also sets moving three most curious little bones. The first of the bones is fastened to the middle of the drumhead so that it moves to and fro every time this membrane quivers. The head of this bone fits into a hole in the next bone, the anvil, and is fastened to it by muscles, so as to drag it along with it; but, the muscles being elastic, it can draw back a little from the anvil, and so give it a blow each time it comes back. This anvil is in its turn very firmly fixed to the little bone, shaped like a stirrup, which you see at the end of the chain.

This stirrup rests upon a curious body which looks in the diagram like a snail-shell with tubes coming out of it. This body, which is called the labyrinth, is made of bone, but it has two little windows in it, one covered only by a membrane, while the other has the head of the stirrup resting upon it.

Now, with a little attention you will understand that when the air in the canal shakes the drumhead to and fro, this membrane must drag with it the hammer, the anvil, and the stirrup. Each time the drum goes in, the hammer will hit the anvil, and drive the stirrup against the little window; every time it goes out it will draw the hammer, the anvil, and the stirrup out again, ready for another blow. Thus the stirrup is always playing upon this little window. Meanwhile, inside the bony labyrinth there is a fluid like water, and along the little passages are very fine hairs, which wave to and fro like reeds; and whenever the stirrup hits at the little window, the fluid moves these hairs to and fro, and they irritate the ends of a nerve, and this nerve carries the message to your brain. There are also some curious little stones called otoliths, lying in some parts of this fluid, and they, by their rolling to and fro, probably keep up the motion and prolong the sound.

You must not imagine we have explained here the many intricacies which occur in the ear; I can only hope to give you a rough idea of it, so that you may picture to yourselves the air-waves moving backwards and forward in the canal of your ear, then the tympanum vibrating to and fro, the hammer hitting the anvil, the stirrup knocking at the little window, the fluid waving the fine hairs and rolling the tiny stones, the ends of the nerve quivering, and then (how we know not) the brain hearing the message.

Is not this wonderful, going on as it does at every sound you hear? And yet his is not all, for inside that curled part of the labyrinth, which looks like a snail-shell and is called the cochlea, there is a most wonderful apparatus of more than three thousand fine stretched filaments or threads, and these act like the strings of a harp, and make you hear different tones. If you go near to a harp or a piano, and sing any particular note very loudly, you will hear this note sounding in the instrument, because you will set just that particular string quivering, which gives the note you sang. The air-waves set going by your voice touch that string, because it can quiver in time with them, while none of the other strings can do so. Now, just in the same way the tiny instrument of three thousand strings in your ear, which is called Corti's organ, vibrates to the air-waves, one thread to one set of waves, and another to another, and according to the fibre that quivers, will be the sound you hear. Here then at last, we see how nature speaks to us. All the movements going on outside, however violent and varied they may be, cannot of themselves make sound. But here, in the little space behind the drum of our ear, the air-waves are sorted and sent on to our brain, where they speak to us as sound.

Week 18

But why then do we not hear all sounds as music? Why are some mere noise, and others clear musical notes? This depends entirely upon whether the sound-waves come quickly and regularly, or by an irregular succession of shocks. For example, when a load of stones is being shot out of a cart, you hear only a long, continuous noise, because the stones fall irregularly, some quicker, some slower, here a number together, and there two or three stragglers by themselves; each of these different shocks comes to your ear and makes a confused, noisy sound. But if you run a stick very quickly along a paling, you will hear a sound very like a musical not. This is because the rods of the paling are all at equal distances one from another, and so the shocks fall quickly one after another at regular intervals upon your ear. Any quick and regular succession of sounds makes a note, even though it may be an ugly one. The squeak of a slate pencil along a slate, and the shriek of a railway whistle are not pleasant, but they are real notes which you could copy on a violin.

I have here a simple apparatus which I have had made to show you that rapid and regular shocks produce a natural musical note. This wheel (Fig. 34) is milled at the edge like a shilling, and when I turn it rapidly so that it strikes against the edge of the card fixed behind it, the notches strike in rapid succession, and produce a musical sound. We can also prove by this experiment that the quicker the blows are, the higher the note will be. I pull the string gently at first, and then quicker and quicker, and you will notice that the note grows sharper and sharper, till the movement begins to slacken, when the note goes down again. This is because the more rapidly the air is hit, the shorter are the waves it makes, and short waves give a high note.

Let us examine this with two tuning-forks. I strike one, and it sounds D, the third space in the treble; I strike the other, and it sounds G, the first leger line, five notes above the C. I have drawn on this diagram (Fig. 35), an imaginary picture of these two sets of waves. You see that the G fork makes three waves, while the C fork makes only two. Why is this? Because the prong of the G fork moves three times backwards and forwards while the prong of the C fork only moves twice; therefore the G fork does not crowd so many atoms together before it draws back, and the waves are shorter. These two notes, C and G, are a fifth of an octave apart; if we had two forks, of which one went twice as fast as the other, making four waves while the other made two, then that note would be an octave higher.

So we see that all the sounds we hear, - the warning noises which keep us from harm, the beautiful musical notes with all the tunes and harmonies that delight us, even the power of hearing the voices of those we love, and learning from one another that which each can tell, - all these depend upon the invisible waves of air, even as the pleasures of light depend on the waves of ether. It is by these sound-waves that nature speaks to us, and in all her movements there is a reason why her boice is sharp or tender, loud or gentle, awful or loving. Take for instance the brook we spoke of at the beginning of the lecture. Why does it sing so sweetly, while the wide deep river makes no noise? Because the little brook eddies and purls round the stones, hitting them as it passes; sometimes the water falls down a large stone, and strikes against the water below; or sometimes it grates the little pebbles together as they lie in its bed. Each of these blows makes a small globe of sound-waves, which spread and spread till they fall on your ear, and because they fall quickly and regularly, they make a low, musical note. We might almost fancy that the brook wished to show how joyfully it flows along, recalling Shelley's beautiful lines:-

"Sometimes it fell
Among the moss with hollow harmony,
Dark and profound; now on the polished stones
It danced; like childhood laughing as it went."

The broad deep river, on the contrary, makes none of these cascades and commotions. The only places against which it rubs are the banks and the bottom; and here you can sometimes hear it grating the particles of sand against each other if you listen very carefully. But there is another reason why falling water makes a sound, and often even a loud roaring noise in the cataract and in the breaking waves of the sea. You do not only hear the water dashing against the rocky ledges or on the beach, you also hear the bursting of innumerable little bladders of air which are contained in the water. As each of these bladders is dashed on the ground, it explodes and sends sound-waves to your ear. Listen to the sea some day when the waves are high and stormy, and you cannot fail to be struck by the irregular bursts of sound.

The waves, however, do not only roar as they dash on the ground; have you never noticed how they seem to scream as they draw back down the beach? Tennyson calls it,

"The scream of the madden'd beach dragged down by the wave;" and it is caused by the stones grating against each other as the waves drag them down. Dr. Tyndall tells us that it is possible to know the size of the stones by the kind of noise they make. If they are large, it is a confused noise, when smaller, a kind of scream; while a gravelly beach will produce a mere hiss.

Who could be dull by the side of a brook, a waterfall, or the sea, while he can listen for sounds like these, and picture to himself how they are being made? You may discover a number of other causes of sound made by water, if you once pay attention to them.

Nor is it only water that sings to us. Listen to the wind, how sweetly it sighs among the leaves. There we hear it, because it rubs the leaves together, and they produce the sound-waves. But walk against the wind some day and you can hear it whistling in your own ear, striking against the curved cup, and then setting up a succession of waves in the hearing canal of the ear itself.

Why should it sound in one particular tone when all kinds of sound-waves must be surging about in the disturbed air?

This glass jar will answer our question roughly. If I strike my tuning-fork and hold it over the jar, you cannot hear it, because the sound is feeble, but if I fill the jar gently with water, when the water rises to a certain point you will hear a loud clear note, because the waves of air in the jar are exactly the right length to answer to the note of the fork. If I now blow across the mouth of the jar you hear the same note, showing that a cavity of a particular length will only sound to the waves which fit it. do you see now the reason why pan-pipes give different sounds, or even the hole at the end of a common key when you blow across it? Here is a subject you will find very interesting if you will read about it, for I can only just suggest it to you here. But now you will see that the canal of your ear also answers only to certain waves, and so the wind sings in your ear with a real if not a musical note.

Again, on a windy night have you not heard the wind sounding a wild, sad note down a valley? Why do you think it sounds so much louder and more musical here than when it is blowing across the plain? Because air in the valley will only answer to a certain set of waves, and, like the pan-pipe, gives a particular note as the wind blows across it, and these waves go up and down the valley in regular pulses, making a wild howl. You may hear the same in the chimney, or in the keyhole; all these are waves set up in the hole across which the wind blows. Even the music in the shell which you hold to your ear is made by the air in the shell pulsating to and fro. And how do you think it is set going? By the throbbing of the veins in your own ear, which causes the air in the shell to vibrate.

Another grand voice of nature is the thunder. People often have a vague idea that thunder is produced by the clouds knocking together, which is very absurd, if you remember that clouds are but water-dust. The most probable explanation of thunder is much more beautiful than this. You will remember from Lecture III that heat forces the air-atoms apart. Now, when a flash of lightning crosses the sky it suddenly expands the air all round it as it passes, so that globe after globe of sound-waves is formed at every point across which the lightning travels. Now light, you remember, travels so wonderfully rapidly (192,000 miles in a second) that a flash of lightning is seen by us and is over in a second, even when it is two or three miles long. But sound comes slowly, taking five seconds to travel half a mile, and so all the sound-waves at each point of the two or three miles fall on our ear one after the other, and make the rolling thunder. Sometimes the roll is made even longer by the echo, as the sound-waves are reflected to and fro by the clouds on their road; and in the mountains we know how the peals echo and re-echo till they die away.

We might fill up far more than an hour in speaking of those voices which come to us as nature is at work. Think of the patter of the rain, how each drop as it hits the pavement sends circles of sound-waves out on all sides; or the loud report which falls on the ear of the Alpine traveller as the glacier cracks on its way down the valley; or the mighty boom of the avalanche as the snow slides in huge masses off the side of the lofty mountain. Each and all of these create their sound-waves, large or small, loud or feeble, which make their way to your ear, and become converted into sound.

We have, however, only time now just to glance at life-sounds, of which there are so many around us. Do you know why we hear a buzzing, as the gnat, the bee, or the cockchafer fly past? Not by the beating of their wings against the air, as many people imagine, and as is really the case with humming birds, but by the scraping of the under-part of their hard wings against the edges of their hind legs, which are toothed like a saw. The more rapidly their wings move the stronger the grating sound becomes, and you will now see why in hot, thirsty weather the buzzing of the gnat is so loud, for the more thirsty and the more eager he

becomes, the wilder his movements will be.
Some insects, like the drone-fly (Eristalis tenax), force the air through the tiny air-passages in their sides, and as these passages are closed by little plates, the plates vibrate to and fro and make sound-waves. Again, what are those curious sounds you may hear sometimes if you rest your head on a trunk in the forest? They are made by the timber-boring beetles, which saw the wood with their jaws and make a noise in the world, even though they have no voice.
All these life-sounds are made by creatures which do not sing or speak; but the sweetest sounds of all in the woods are the voices of the birds. All voice-sounds are made by two elastic bands or cushions, called vocal chords, stretched across the end of the tube or windpipe through which we breathe, and as we send the air through them we tighten or loosen them as we will, and so make them vibrate quickly or slowly and make sound-waves of different lengths. But if you will try some day in the woods you will find that a bird can beat you over and over again in the length of his note; when you are out of breath and forced to stop he will go on with his merry trill as fresh and clear as if he had only just begun. This is because birds can draw air into the whole of their body, and they have a large stock laid up in the folds of their windpipe, and besides this the air-chamber behind their elastic bands or vocal chords has two compartments where we have only one, and the second compartment has special muscles by which they can open and shut it, and so prolong the trill.
Only think what a rapid succession of waves must quiver through the air as a tiny lark agitates his little throat and pours forth a volume of song! The next time you are in the country in the spring, spend half an hour listening to him, and try and picture to yourself how that little being is moving all the atmosphere round him. Then dream for a little while about sound, what it is, how marvellously it works outside in the world, and inside in your ear and brain; and then, when you go back to work again, you will hardly deny that it is well worth while to listen sometimes to the voices of nature and ponder how it is that we hear them.

Week 19

LECTURE VII THE LIFE OF A PRIMROSE

When the dreary days of winter and the early damp days of spring are passing away, and the warm bright sunshine has begun to pour down upon the grassy paths of the wood, who does not love to go out and bring home posies of violets, and bluebells, and primroses? We wander from one plant to another picking a flower here and a bud there, as they nestle among the green leaves, and we make our rooms sweet and gay with the tender and lovely blossoms. But tell me, did you ever stop to think, as you added flower after flower to your nosegay, how the plants which bear them have been building up their green leaves and their fragile buds during the last few weeks? If you had visited the same spot a month before, a few (of) last year's leaves, withered and dead, would have been all that you would have found. And now the whole wood is carpeted with delicate green leaves, with nodding bluebells, and pale-yellow primroses, as if a fairy had touched the ground and covered it with fresh young life. And our fairies have been at work here; the fairy "Life," of whom we know so little, though we love her so well and rejoice in the beautiful forms she can produce; the fairy sunbeams with their invisible influence kissing the tiny shoots and warming them into vigour and activity; the gentle rain-drops, the balmy air, all these have been working, while you or I passed heedlessly by; and now we come and gather the flowers they have made, and too often forget to wonder how these lovely forms have sprung up around us.

Our work during the next hour will be to consider this question. You were asked last week to bring with you to-day a primrose- flower, or a whole plant if possible, in order the better to follow out with me the "Life of a Primrose." (To enjoy this lecture, the reader ought to have, if possible, a primrose- flower, an almond soaked for a few minutes in hot water, and a piece of orange.) This is a very different kind of subject from those of our former lectures. There we took world- wide histories; we travelled up to the sun, or round the earth, or into the air; now I only ask you to fix your attention on one little plant, and inquire into its history.

There is a beautiful little poem by Tennyson, which says -

"Flower in the crannied wall,
I pluck you out of the crannies;
Hold you here, root and all, in my hand,
Little flower; but if I could understand
What you are, root and all, and all in all,
I should know what God and man is."

We cannot learn all about this little flower, but we can learn enough to understand that it has a real separate life of its own, well worth knowing. For a plant is born, breathes, sleeps, feeds, and digests just as truly as an animal does, though in a different way. It works hard both for itself to get its food, and for others in making the air pure and fit for animals to breathe. It often lays by provision for the winter. It sends young plants out, as parents send their children, to fight for themselves in the world; and then, after living sometimes to a good old age, it dies, and leaves its place to others.

We will try to follow out something of this life to-day; and first, we will begin with the seed.

I have here a packet of primrose-seeds, but they are so small that we cannot examine them; so I have also had given to each one of you an almond-kernel, which is the seed of the almond- tree, and which has been soaked, so that it splits in half easily. From this we can learn about seeds in general, and then apply it to the primrose.

If you peel the two skins off your almond-seed (the thick, brown, outside skin, and the thin, transparent one under it), the two halves of the almond will slip apart quite easily. One of these halves will have a small dent at the pointed end, while in the other half you will see a little lump, which fitted into the dent when the two halves were joined. This little lump (a b, Fig. 37) is a young plant, and the two halves of the almond are the seed leaves which hold the plantlet, and feed it till it can feed itself. The rounded end of the plantlet (b) sticking out of the almond, is the beginning of the root, while the other end (a) will in time become the stem. If you look carefully, you will see two little points at this end, which are the tips of future leaves. Only think how minute this plantlet must be in a primrose, where the whole seed is scarcely larger than a grain of sand! Yet in this tiny plantlet lies hid the life of the future plant.

When a seed falls into the ground, so long as the earth is cold and dry, it lies like a person in a trance, as if it were dead; but as soon as the warm, damp spring comes, and the busy little sun-waves pierce down into the earth, they wake up the plantlet and make it bestir itself. They agitate to and fro the particles of matter in this tiny body, and cause them to seek out for other particles to seize and join to themselves.

But these new particles cannot come in at the roots, for the seed has none; nor through the leaves, for they have not yet grown up; and so the plantlet begins by helping itself to the store of food laid up in the thick seed-leaves in which it is buried. Here it finds starch, oils, sugar, and substances called albuminoids, — the sticky matter which you notice in wheat-grains when you chew them is one of the albuminoids. This food is all ready for the plantlet to use, and it sucks it in, and works itself into a young plant with tiny roots at one end, and a growing shoot, with leaves, at the other.

But how does it grow? What makes it become larger? To answer this you must look at the second thing I asked you to bring - a piece of orange. If you take the skin off a piece of orange, you will see inside a number of long-shaped transparent bags, full of juice. These we call cells, and the flesh of all plants and animals is made up of cells like these, only of various shapes. In the pith of elder they are round, large, and easily seen (a, Fig. 39); in the stalks of plants they are long, and lap over each other (b, Fig. 39), so as to give the stalk strength to stand upright. Sometimes many cells growing one on the top of the other break into one tube and make vessels. But whether large or small, they are all bags growing one against the other.

In the orange-pulp these cells contain only sweet juice, but in other parts of the orange-tree or any other plant they contain a sticky substance with little grains in it. This substance is called "protoplasm," or the first form of life, for it is alive and active, and under a microscope you may see in a living plant streams of the little grains moving about in the cells.

Now we are prepared to explain how our plant grows. Imagine the tiny primrose plantlet to be made up of cells filled with active living protoplasm, which drinks in starch and other food from the seed-leaves. In this way each cell will grow too full for its skin, and then the protoplasm divides into two parts and builds up a wall between them, and so one cell becomes two. Each of these two cells again breaks up into two more, and so the plant grows larger and larger, till by the time it has used up all the food in the seed-leaves, it has sent roots covered with fine hairs downwards into the earth, and a shoot with beginnings of leaves up into the air.

Sometimes the seed-leaves themselves come above the ground, as in the mustard-plant, and sometimes they are left empty behind, while the plantlet shoots through them.

And now the plant can no longer afford to be idle and live on prepared food. It must work for itself. Until now it has been taking in the same kind of food that you and I do; for we too find many seeds very pleasant to eat and useful to nourish us. But now this store is exhausted. Upon what then is the plant to live? It is cleverer than we are in this, for while we cannot live unless we have food which has once been alive, plants can feed upon gases and water and mineral matter only. Think over the substances you can eat or drink, and you will find they are nearly all made of things which have been alive: meat, vegetables, bread, beer, wine, milk; all these are made from living matter, and though you do take in such things as water and salt, and even iron and phosphorus, these would be quite useless if you did not eat and drink prepared food which your body can work into living matter.

But the plant as soon as it has roots and leaves begins to make living matter out of matter that has never been alive. Through all the little hairs of its roots it sucks in water, and in this water are dissolved more or less of the salts of ammonia, phosphorus, sulphur, iron, lime, magnesia, and even silica, or flint. In all kinds of earth there is some iron, and we shall see presently that this is very important to the plant.

Suppose, then, that our primrose has begun to drink in water at its roots. How is it to get this water up into the stem and leaves, seeing that the whole plant is made of closed bags or cells? It does it in a very curious way, which you can prove for yourselves. Whenever two fluids, one thicker than the other, such as treacle and water for example, are only separated by a skin or any porous substance, they will always mix, the thinner one oozing through the skin into the thicker one. If you tie a piece of bladder over a glass tube, fill the tube half-full of treacle, and then let the covered end rest in a bottle of water, in a few hours the water will get in to the treacle and the mixture will rise up in the tube till it flows over the top. Now, the saps and juices of plants are thicker than water, so, directly the water enters the cells at the root it oozes up into the cells above, and mixes with the sap. Then the matter in those cells becomes thinner than in the cells above, so it too oozes up, and in this way cell by cell the water is pumped up into the leaves.

When it gets there it finds our old friends the sun-beams hard at work. If you have ever tried to grow a plant in a cellar, you will know that in the dark its leaves remain white and sickly. It is only in the sunlight that a beautiful delicate green tint is given to them, and you will remember from Lecture II. that this green tint shows that the leaf has used all the sun-waves except those which make you see green; but why should it do this only when it has grown up in the sunshine?

The reason is this: when the sunbeam darts into the leaf and sets all its particles quivering, it divides the protoplasm into two kinds, collected into different cells. One of these remains white, but the other kind, near the surface, is altered by the sunlight and by the help of the iron brought in by the water. This particular kind of protoplasm, which is called "chlorophyll," will have nothing to do with the green waves and throws them back, so that every little grain of this protoplasm looks green and gives the leaf its green colour.

It is these little green cells that by the help of the sun-waves digest the food of the plant and turn the water and gases into useful sap and juices. We saw in Lecture III. that when we breathe-in air, we use up the oxygen in it and send back out of our mouths carbonic acid, which is a gas made of oxygen and carbon.

Now, every living things wants carbon to feed upon, but plants cannot take it in by itself, because carbon is solid (the blacklead in your pencils is pure carbon), and a plant cannot eat, it can only drink-in fluids and gases. Here the little green cells help it out of its difficulty. They take in or absorb out of the air carbonic acid gas which we have given out of our mouths and then by the help of the sun-waves they tear the carbon and oxygen apart. Most of the oxygen they throw back into the air for us to use, but the carbon they keep.

If you will take some fresh laurel-leaves and put them into a tumbler of water turned upside-down in a saucer of water, and set the tumbler in the sunshine, you will soon see little bright bubbles rising up and clinging to the glass. These are bubbles of oxygen gas, and they tell you that they have been set free by the green cells which have torn from them the carbon of the carbonic acid in the water.

But what becomes of the carbon? And what use is made of the water which we have kept waiting all this time in the leaves? Water, you already know, is made of hydrogen and oxygen, but perhaps you will be surprised when I tell you that starch, sugar, and oil, which we get from plants, are nothing more than hydrogen and oxygen in different quantities joined to carbon.

It is very difficult at first to picture such a black thing as carbon making part of delicate leaves and beautiful flowers, and still more of pure white sugar. But we can make an experiment by which we can draw the hydrogen and oxygen out of common loaf sugar, and then you will see the carbon stand out in all its blackness. I have here a plate with a heap of white sugar in it. I pour upon it first some hot water to melt and warm it, and then some strong sulphuric acid. This acid does nothing more than simply draw the hydrogen and oxygen out. See! in a few moments a black mass of carbon begins to rise, all of which has come out of the white sugar you saw just now. *(The common dilute sulphuric acid of commerce is not strong enough for this experiment, but pure sulphuric acid can be secured from any chemist. Great care must be taken in using it, as it burns everything it touches.) You see, then, that from the whitest substance in plants we can get this black carbon; and in truth, one-half of the dry part of every plant is composed of it.

Now look at my plant again, and tell me if we have not already found a curious history? Fancy that you see the water creeping in at the roots, oozing up from cell to cell till it reaches the leaves, and there meeting the carbon which has just come out of the air, and being worked up with it by the sun-waves into starch, or sugar, or oils.

But meanwhile, how is new protoplasm to be formed? for without this active substance none of the work can go on. Here comes into use a lazy gas we spoke of in Lecture III. There we thought that nitrogen was of no use except to float oxygen in the air, but here we shall find it very useful. So far, as we know, plants cannot take up nitrogen out of the air, but they

can get it out of the ammonia which the water brings in at their roots.
Ammonia, you will remember, is a strong-smelling gas, made of hydrogen and nitrogen, and which is often almost stifling near a manure-heap. When you manure a plant you help it to get this ammonia, but at any time it gets some from the soil and also from the rain-drops which bring it down in the air. Out of this ammonia the plant takes the nitrogen and works it up with the three elements, carbon, oxygen, and hydrogen, to make the substances called albuminoids, which form a large part of the food of the plant, and it is these albuminoids which go to make protoplasm. You will notice that while the starch and other substances are only made of three elements, the active protoplasm is made of these three added to a fourth, nitrogen, and it also contains phosphorus and sulphur.

And so hour after hour and day after day our primrose goes on pumping up water and ammonia from its roots to its leaves, drinking in carbonic acid from the air, and using the sun-waves to work them all up into food to be sent to all parts of its body. In this way these leaves act, you see, as the stomach of the plant, and digest its food.

Sometimes more water is drawn up into the leaves than can be used, and then the leaf opens thousands of little mouths in the skin of its under surface, which let the drops out just as drops of perspiration ooze through our skin when we are overheated. These little mouths, which are called stomates (a, Fig. 42) are made of two flattened cells, fitting against each other. When the air is damp and the plant has too much water these lie open and let it out, but when the air is dry, and the plant wants to keep as much water as it can, then they are closely shut. There are as many as a hundred thousand of these mouths under one apple-leaf, so you may imagine how small they often are.

Plants which only live one year, such as mignonette, the sweet pea, and the poppy, take in just enough food to supply their daily wants and to make the seeds we shall speak of presently. Then, as soon as their seeds are ripe their roots begin to shrivel, and water is no longer carried up. The green cells can no longer get food to digest, and they themselves are broken up by the sunbeams and turn yellow, and the plant dies.

But many plants are more industrious than the stock and mignonette, and lay by store for another year, and our primrose is one of these. Look at this thick solid mass below the primrose leaves, out of which the roots spring. (See the plant in the foreground of the heading of the lecture.) This is really the stem of the primrose hidden underground, and all the starch, albuminoids, &c., which the plant can spare as it grows, are sent down into this underground stem and stored up there, to lie quietly in the ground through the long winter, and then when the warm spring comes this stem begins to send out leaves for a new plant.

Week 21

We have now seen how a plant springs up, feeds itself, grows, stores up food, withers, and dies; but we have said nothing yet about its beautiful flowers or how it forms its seeds. If we look down close to the bottom of the leaves in a primrose root in spring-time, we shall always find three or four little green buds nestling in among the leaves, and day by day we may see the stalk of these buds lengthening till they reach up into the open sunshine, and then the flower opens and shows its beautiful pale-yellow crown.

We all know that seeds are formed in the flower, and that the seeds are necessary to grow into new plants. But do we know the history of how they are formed, or what is the use of the different parts of the bud? Let us examine them all, and then I think you will agree with me that this is not the least wonderful part of the plant.

Remember that the seed is the one important thing and then notice how the flower protects it. First, look at the outside green covering, which we call the calyx. See how closely it fits in the bud, so that no insects can creep in to gnaw the flower, nor any harm come to it from cold or blight. Then, when the calyx opens, notice that the yellow leaves which form the crown or corolla, are each alternate with one of the calyx leaves, so that anything which got past the first covering would be stopped by the second. Lastly, when the delicate corolla has opened out, look at those curious yellow bags just at the top of the tube (b,2, Fig. 43). What is their use?

But I fancy I see two or three little questioning faces which seem to say, "I see no yellow bags at the top of the tube." Well, I cannot tell whether you can or not in the specimen you have in your hand; for one of the most curious things about primrose flowers is, that some of them have these yellow bags at the top of the tube and some of them hidden down right in the middle. But this I can tell you:those of you who have got no yellow bags at the top will have a round knob there (I a, Fig. 43), and will find the yellow bags (b) buried in the tube. Those, on the other hand, who have the yellow bags (2 b, Fig. 43) at the top will find the knob (a) half-way down the tube.

Now for the use of these yellow bags, which are called the anthers of the stamens, the stalk on which they grow being called the filament or thread. If you can manage to split them open you will find that they have a yellow powder in them, called pollen, the same as the powder which

sticks to your nose when you put it into a lily; and if you look with a magnifying glass at the little green knob in the centre of the flower, you will probably see some of this yellow dust sticking on it (A, Fig. 43). We will leave it there for a time, and examine the body called the pistil, to which the knob belongs. Pull off the yellow corolla (which will come off quite easily), and turn back the green leaves. You will then see that the knob stands on the top of a column, and at the bottom of this column there is a round ball (s v), which is a vessel for holding the seeds. In this diagram (A, Fig. 43) I have drawn the whole of this curious ball and column as if cut in half, so that we may see what is in it. In the middle of the ball, in a cluster, there are a number of round transparent little bodies, looking something like round green orange-cells full of juice. They are really cells full of protoplasm, with one little dark spot in each of them, which by-and-by is to make our little plantlet that we found in the seed.
"These, then, are seeds," you will say. Not yet; they are only ovules, or little bodies which may become seeds. If they were left as they are they would all wither and die. But those little grains of pollen, which we saw sticking to the knob at the top, are coming down to help them. As soon as these yellow grains touch the sticky knob or stigma, as it is called, they throw out tubes, which grow down the column until they reach the ovules. In each one of these they find a tiny hole, and into this they creep, and then they pour into the ovule all the protoplasm from the pollen-grain which is sticking above, and this enables it to grow into a real seed, with a tiny plantlet inside.
This is how the plant forms its seed to bring up new little ones next year, while the leaves and the roots are at work preparing the necessary food. Think sometimes when you walk in the woods, how hard at work the little plants and big trees are, all around you. You breathe in the nice fresh oxygen they have been throwing out, and little think that it is they who are making the country so fresh and pleasant, and that while they look as if they were doing nothing but enjoying the bright sunshine, they are really fulfilling their part in the world by the help of this sunshine; earning their food from the ground working it up; turning their leaves where they can best get light (and in this it is chiefly the violet sun-waves that help them), growing, even at night, by making new cells out of the food they have taken in the day; storing up for the winter; putting out their flowers and making their seeds, and all the while smiling so pleasantly in quiet nooks and sunny dells that it makes us glad to see them.
But why should the primroses have such golden crowns? plain green ones would protect the seed quite as well. Ah! now we come to a secret well worth knowing. Look at the two primrose flowers, 1 and 2, Fig. 43,

p. 163, and tell me how you think the dust gets on to the top of the sticky knob or stigma. No. 2 seems easy enough to explain, for it looks as if the pollen could fall down easily from the stamens on to the knob, but it cannot fall up, as it would have to do in No. 1. Now the curious truth is, as Mr. Darwin has shown, that neither of these flowers can get the dust easily for themselves, but of the two No. 1 has the least difficulty.
Look at a withered primrose, and see how it holds its head down, and after a little while the yellow crown falls off. It is just about as it is falling that the anthers or bags of stamens burst open, and then, in No. 1 (Fig. 44), they are dragged over the knob and some of the grains stick there. But in the other form of primrose, No. 2, when the flower falls off, the stamens do not come near the knob, so it has no chance of getting any pollen; and while the primrose is upright the tube is so narrow that the dust does not easily fall. But, as I have said, neither kind gets it very easily, nor is it good for them if they do. The seeds are much stronger and better if the dust or pollen of one flower is carried away and left on the knob or stigma of another flower; and the only way this can be done is by insects flying from one flower to another and carrying the dust on their legs and bodies.
If you suck the end of the tube of the primrose flower you will find it tastes sweet, because a drop of honey has been lying there. When the insects go in to get this honey, they brush themselves against the yellow dust-bags, and some of the dust sticks to them, and then when they go to the next flower they rub it off on to its sticky knob.
Look at No. 1 and No. 2 (Fig. 43) and you will see at once that if an insect goes into No. 1 and the pollen sticks to him, when he goes into No. 2 just that part of his body on which the pollen is will touch the knob; and so the flowers become what we call "crossed," that is, the pollen-dust of the one feeds the ovule of the other. And just the same thing will happen if he flies from No. 2 to No. 1. There the dust will be just in the position to touch the knob which sticks out of the flower.
Therefore, we can see clearly that it is good for the primrose that bees and other insects should come to it, and anything it can do to entice them will be useful. Now, do you not think that when an insect once knew that the pale-yellow crown showed where honey was to be found, he would soon spy these crowns out as he flew along? or if they were behind a hedge, and he could not see them, would not the sweet scent tell him where to come and look for them? And so we see that the pretty sweet-scented corolla is not only delightful for us to look at and to smell, but it is really very useful in helping the primrose to make strong healthy seeds out of which the young plants are to grow next year.
And now let us see what we have learnt. We began with a tiny seed,

though we did not then know how this seed had been made. We saw the plantlet buried in it, and learnt how it fed at first on prepared food, but soon began to make living matter for itself out of gases taken from the water through the cells to its stomach - the leaves! And how marvellously the sun-waves entering there formed the little green granules, and then helped them to make food and living protoplasm! At this point we might have gone further, and studied how the fibres and all the different vessels of the plant are formed, and a wondrous history it would have been. But it was too long for one hour's lecture, and you must read it for yourselves in books on botany. We had to pass on to the flower, and learn the use of the covering leaves, the gaily coloured crown attracting the insects, the dust-bags holding the pollen, the little ovules each with the germ of a new plantlet, lying hidden in the seed-vessel, waiting for the pollen-grains to grow down to them. Lastly, when the pollen crept in at the tiny opening we learnt that the ovule had now all it wanted to grow into a perfect seed.

And so we came back to a primrose seed, the point from which we started; and we have a history of our primrose from its birth to the day when its leaves and flowers wither away and it dies down for the winter.

But what fairies are they which have been at work here? First, the busy little fairy Life in the active protoplasm; and secondly, the sun-waves. We have seen that it was by the help of the sunbeams that the green granules were made, and the water, carbonic acid, and nitrogen worked up into the living plant. And in doing this work the sun-waves were caught and their strength used up, so that they could no longer quiver back into space. But are they gone for ever? So long as the leaves or the stem or the root of the plant remain they are gone, but when those are destroyed we can get them back again. Take a handful of dry withered plants and light them with a match, then as the leaves burn and are turned back again to carbonic acid, nitrogen, and water, our sunbeams come back again in the flame and heat.

And the life of the plant? What is it, and why is this protoplasm always active and busy? I cannot tell you. Study as we may, the life of the tiny plant is as much a mystery as your life and mine. It came, like all things, from the bosom of the Great Father, but we cannot tell how it came nor what it is. We can see the active grains moving under the microscope, but we cannot see the power that moves them. We only know it is a power given to the plant, as to you and to me, to enable it to live its life, and to do its useful work in the world.

Week 22

LECTURE VIII

THE HISTORY OF A PIECE OF COAL

I have here a piece of coal (Fig. 45), which, though it has been cut with some care so as to have a smooth face, is really in no other way different from any ordinary lump which you can pick for yourself out of the coal-scuttle. Our work to-day is to relate the history of this black lump; to learn what it is, what it has been, and what it will be.

It looks uninteresting enough at first sight, and yet if we examine it closely we shall find some questions to ask even about its appearance. Look at the smooth face of this specimen and see if you can explain those fine lines which run across so close together as to look like the edges of the leaves of a book. Try to break a piece of coal, and you will find that it will split much more easily along those lines than across the other way of the lump; and if you wish to light a fire quickly you should always put this lined face downwards so that the heat can force its way up through these cracks and gradually split up the block. Then again if you break the coal carefully along one of these lines you will find a fine film of charcoal lying in the crack, and you will begin to suspect that this black coal must have been built up in very thin layers, with a kind of black dust between them.

The next thing you will call to mind is that this coal burns and gives flame and heat, and that this means that in some way sunbeams are imprisoned in it; lastly, this will lead you to think of plants, and how they work up the strength of the sunbeams into their leaves, and hide black carbon in even the purest and whitest substance they contain.

Is coal made of burnt plants, then? Not burnt ones, for if so it would not burn again; but you may have read how the makers of charcoal take wood and bake it without letting it burn, and then it turns black and will afterwards make a very good fire; and so you will see that it is probable that our piece of coal is made of plants which have been baked and altered, but which have still much sunbeam strength bottled up in them, which can be set free as they burn.

If you will take an imaginary journey with me to a coal-pit near Newcastle, which I visited many years ago, you will see that we have very good evidence that coal is made of plants, for in all coal-mines we find remains of them at every step we take.

Let us imagine that we have put on old clothes which will not spoil, and have stepped into the iron basket (see Fig. 46) called by the miners a cage, and are being let down the shaft to the gallery where the miners are at work. Most of them will probably be in the gallery b, because a great deal of the coal in a has been already taken out. But we will stop in a because there we can see a great deal of the roof and the floor. When we land on the floor of the gallery we shall find ourselves in a kind of tunnel with railway lines laid along it and trucks laden with coal coming towards the cage to be drawn up, while empty ones are running back to be loaded where the miners are at work. Taking lamps in our hands and keeping out of the way of the trucks, we will first throw the light on the roof, which is made of shale or hardened clay. We shall not have gone many yards before we see impressions of plants in the shale, like those in this specimen (Fig. 47), which was taken out of a coal-mine at Neath in Glamorganshire, a few days ago, and sent up for this lecture. You will recognize at once the marks of ferns (a), for they look like those you gather in the hedges of an ordinary country lane, and that long striped branch (b) does not look unlike a reed, and indeed it is something of this kind, as we shall see by-and-by. You will find plenty of these impressions of plants as you go along the gallery and look up at the roof, and with them there will be others with spotted stems, or with stems having a curious diamond pattern upon them, and many ferns of various kinds.

Next look down at your feet and examine the floor. You will not have to search long before you will almost certainly find a piece of stone like that represented in Fig. 48, which has also come from Neath Colliery. This fossil, which is the cast of a piece of a plant, puzzled those who found it for a very long time. At last, however, Mr. Binney found the specimen growing to the bottom of the trunk of one of the fossil trees with spotted stems, called Sigillaria; and so proved that this curious pitted stone is a piece of fossil root, or rather underground stem, like that which we found in the primrose, and that the little pits or dents in it are scars where the rootlets once were given off.

Whole masses of these root-stems, with ribbon-like roots lying scattered near them, are found buried in the layer of clay called the underclay which makes the floor of the coal, and they prove to us that this underclay must have been once the ground in which the roots of the coal-plants grew. You will feel still more sure of this when you find that there is not only one straight gallery of coal, but that galleries branch out right and left, and that everywhere you find the coal lying like a sandwich between the floor and the roof, showing that quite a large piece of country must be covered by these remains of plants all rooted in the underclay.

But how about the coal itself? It seems likely, when we find roots below and leaves and stems above, that the middle is made of plants, but can we prove it? We shall see presently that it has been so crushed and altered by being buried deep in the ground that the traces of leaves have almost been destroyed, though people who are used to examining with the microscope, can see the crushed remains of plants in thin slices of coal.

But fortunately for us, perfect pieces of plants have been preserved even in the coal-bed itself. Do you remember our learning in Lecture IV, that water with lime in it petrifies things, that is, leaves carbonate of lime to fill up grain by grain the fibres of an animal or plant as the living matter decays, and so keeps an exact representation of the object?

Now, it so happens that in a coal-bed at South Ouram, near Halifax, as well as in some other places, carbonate of lime trickled in before the plants were turned into coal, and made some round nodules in the plant-bed, which look like cannon- balls. Afterwards, when all the rest of the bed was turned into coal, these round balls remained crystallized, and by cutting thin transparent slices across the nodule we can distinctly see the leaves and stems and curious little round bodies which make up the coal. Several such sections may be seen at the British Museum, and when we compare these fragments of plants with those which we find above and below the coal-bed, we find that they agree, thus proving that coal is made of plants, and of those plants whose roots grew in the clay floor, while their heads reached up far above where the roof now is.

The next question is, what kind of plants were these? Have we anything like them living in the world now? You might perhaps think that it would be impossible to decide this question from mere petrified pieces of plants. But many men have spent their whole lives in deciphering all the fragments that could be found, and though the section given in Fig. 49 may look to you quite incomprehensible, yet a botanist can reed it as we read a book. For example, at S and L, where stems are cut across, he can learn exactly how they were build up inside, and compare them with the stems of living plants, while the fruits cc and the little round spores lying near them, tell him their history as well as if he had gathered them from the tree. In this way we have learnt to know very fairly what the plants of the coal were like, and you will be surprised when I tell you that the huge trees of the coal-forests, of which we sometimes find trunks in the coal-mines from ten to fifty feet long, are only represented on the earth now by small insignificant plants, scarcely ever more than two feet, and often not many inches high.

Have you ever seen the little club moss or Lycopodium which grows all over England, but chiefly in the north, on heaths and mountains? At the end of each of its branches it bears a cone made of scaly leaves; and fixed to the inside of each of these leaves is a case called a sporangium, full of little spores or moss-seeds, as we may call them, though they are not exactly like true seeds. In one of these club-mosses called Selaginella, the cases near the bottom of the cone contain large spores, while those near the top contain a powdery dust. These spores are full of resin, and they are collected on the Continent for making artificial lightning in the theatres, because they flare when lighted.

Now this little Selaginella is of all living plants the one most like some of the gigantic trees of the coal-forests. If you look at this picture of a coal-forest (Fig. 51), you will find it difficult perhaps to believe that those great trees, with diamond markings all up the trunk, hanging over from the right to the left of the picture, and covering all the top with their boughs, could be in any way relations of the little Selaginella; yet we find branches of them in the beds above the coal, bearing cones larger but just like Selaginella cones; and what is most curious, the spores in these cones are of exactly the same kind and not any larger than those of the club-mosses.

These trees are called by botanists Lepidodendrons, or scaly trees; there are numbers of them in all coal-mines, and one trunk has been found 49 feet long. Their branches were divided in a curious forked manner and bore cones at the ends. The spores which fell from these cones are found flattened in the coal, and they may be seen scattered about in the coal-ball.

Week 23

Another famous tree which grew in the coal-forests was the one whose roots we found in the floor or underclay of the coal. It has been called

Sigillaria, because it has marks like seals (sigillum, a seal) all up the trunk, due to the scars left by the leaves when they fell from the tree. You will see the Sigillarias on the left-hand side of the coal-forest picture, having those curious tufts of leaves springing out of them at the top. Their stems make up a great deal of the coal, and the bark of their trunks is often found in the clays above, squeezed flat in lengths of 30, 60, or 70 feet. Sometimes, instead of being flat the bark is still in the shape of a trunk, and the interior is filled with sane; and then the trunk is very heavy, and if the miners do not prop the roof up well it falls down and kills those beneath it. Stigmaria is the root of the Sigillaria, and is found in the clays below the coal. Botanists are not yet quite certain about the seed-cases of this tree, but Mr. Carruthers believes that they grew inside the base of the leaves, as they do in the quillwort, a small plant which grows at the bottom of our mountain lakes.
But what is that curious reed-like stem we found in the piece of shale (see Fig. 47)? That stem is very important, for it belonged to a plant called a Calamite, which, as we shall see presently, helped to sift the earth away from the coal and keep it pure. This plant was a near relation of the "horsetail," or Equisetum, which grows in our marshes; only, just as in the case of the other trees, it was enormously larger, being often 20 feet high, whereas the little Equisetum, Fig. 52, is seldom more than a foot, and never more than 4 feet high in England, though in tropical South America they are much higher. Still, if you have ever gathered "horsetails," you will see at once that those trees in the foreground of the picture (Fig. 51), with leaves arranged in stars round the branches, are only larger copies of the little marsh-plants; and the seed-vessels of the two plants are almost exactly the same.
These great trees, the Lepidodendrons, the Sigillarias, and the Calamites, together with large tree-ferns, are the chief plants that we know of in the coal-forests. It seems very strange at first that they should have been so large when their descendants are now so small, but if you look at our chief plants and trees now, you will find that nearly all of them bear flowers, and this is a great advantage to them, because it tempts the insects to bring them the pollen-dust, as we saw in the last lecture.
Now the Lipidodendrons and their companions had no true flowers, but only these seed-cases which we have mentioned; but as there were no flowering plants in their time, and they had the ground all to themselves, they grew fine and large. By-and-by, however, when the flowering plants came in, these began to crowd out the old giants of the coal-forests, so that they dwindled and dwindled from century to century till their great-great- grandchildren, thousands of generations after, only lift up their tiny heads in marshes and on heaths, and tell us that they were

big once upon a time.

And indeed they must have been magnificent in those olden days, when they grew thick and tall in the lonely marshes where plants and trees were the chief inhabitants. We find no traces in the clay-beds of the coal to lead us to suppose that men lived in those days, nor lions, nor tigers, nor even birds to fly among the trees; but these grand forests were almost silent, except when a huge animal something like a gigantic newt or frog went croaking through the marsh, or a kind of grasshopper chirruped on the land. But these forms of life were few and far between, compared to the huge trees and tangled masses of ferns and reeds which covered the whole ground, or were reflected in the bosom of the large pools and lakes round about which they grew.

And now, if you have some idea of the plants and trees of the coal, it is time to ask how these plants became buried in the earth and made pure coal, instead of decaying away and leaving behind only a mixture of earth and leaves?

To answer this question, I must ask you to take another journey with me across the Atlantic to the shores of America, and to land at Norfolk in Virginia, because there we can see a state of things something like the marshes of the coal-forests. All round about Norfolk the land is low, flat, and marshy, and to the south of the town, stretching far away into North Carolina, is a large, desolate swamp, no less than forty miles long and twenty-five broad. The whole place is one enormous quagmire, overgrown with water-plants and trees. The soil is as black as ink from the old, dead leaves, grasses, roots, and stems which lie in it; and so soft, that everything would sink into it, if it were not for the matted roots of the mosses, ferns, and other plants which bind it together. You may dig down for ten or fifteen feet, and find nothing but peat made of the remains of plants which have lived and died there in succession for ages and ages, while the black trunks of the fallen trees lie here and there, gradually being covered up by the dead plants.

The whole place is so still, gloomy, and desolate, that it goes by the name of the "Great Dismal Swamp," and you see we have here what might well be the beginning of a bed of coal; for we know that peat when dried becomes firm and makes an excellent fire, and that if it were pressed till it was hard and solid it would not be unlike coal. If, then, we can explain how this peaty bed has been kept pure from earth, we shall be able to understand how a coal-bed may have been formed, even though the plants and trees which grow in this swamp are different from those which grew in the coal-forests.

The explanation is not difficult; streams flow constantly, or rather ooze into the Great Dismal Swamp from the land that lies to the west, but instead of bringing mud in with them as rivers bring to the sea, they bring only clear, pure water, because, as they filter for miles through the dense jungle of reeds, ferns, and shrubs which grow round the marsh, all the earth is sifted out and left behind. In this way the spongy mass of dead plants remains free from earthy grains, while the water and the shade of the thick forest of trees prevent the leaves, stems, etc., from being decomposed by the air and sun. And so year after year as the plants die they leave their remains for other plants to take root in, and the peaty mass grows thicker and thicker, while tall cedar trees and evergreens live and die in these vast, swampy forests, and being in loose ground are easily blown down by the wind, and leave their trunks to be covered up by the growing moss and weeds.

Now we know that there were plenty of ferns and of large Calamites growing thickly together in the coal-forests, for we find their remains everywhere in the clay, so we can easily picture to ourselves how the dense jungle formed by these plants would fringe the coal-swamp, as the present plants do the Great Dismal Swamp, and would keep out all earthy matter, so that year after year the plants would die and form a thick bed of peat, afterwards to become coal.

Week 24

The next thing we have to account for is the bed of shale or hardened clay covering over the coal. Now we know that from time to time land has gone slowly up and down on our globe so as in some places to carry the dry ground under the sea, and in others to raise the sea-bed above the water. Let us suppose, then, that the great Dismal Swamp was gradually to sink down so that the sea washed over it and killed the reeds and shrubs. Then the streams from the west would not be sifted any longer but would bring down mud, and leave it, as in the delta of the Nile or Mississippi, to make a layer over the dead plants. You will easily understand that this mud would have many pieces of dead trees and plants in it, which were stifled and died as it covered them over; and thus the remains would be preserved like those which we find now in the roof of the coal-galleries.

But still there are the thick sandstones in the coal-mine to be explained. How did they come there? To explain them, we must suppose that the ground went on sinking till the sea covered the whole place where once the swamp had been, and then sea-sand would be thrown down over the clay and gradually pressed down by the weight of new sand above, till it formed solid sandstone and our coal-bed became buried deeper and deeper in the earth.

At last, after long ages, when the thick mass of sandstones above the bed b (Fig. 46) had been laid down, the sinking must have stopped and the land have risen a little, so that the sea was driven back; and then the rivers would bring down earth again and make another clay-bed. Then a new forest would spring up, the ferns, Calamites, Lepidodendrons, and Sigillarias would gradually form another jungle, and many hundred of feet above the buried coal-bed b, a second bed of peat and vegetable matter would begin to accumulate to form the coal-bed a.

Such is the history of how the coal which we now dig out of the depths of the earth once grew as beautiful plants on the surface. We cannot tell exactly all the ground over which these forests grew in England, because some of the coal they made has been carried away since by rivers and cut down by the waves of the sea, but we can say that wherever there is coal now, there they must have been then.

Try and picture to yourselves that on the east coast of Northumberland and Durham, where all is now black with coal- dust, and grimy with the smoke of furnaces; and where the noise of hammers and steam-engines, and of carts and trucks hurrying to and fro, makes the country re-echo with the sound of labour; there ages ago in the silent swamp shaded with monster trees, one thin layer of plants after another was formed, year after year, to become the coal we now value so much. In Lancashire, busy Lancashire, the same thing was happening, and even in the middle of Yorkshire and Derbyshire the sea must have come up and washed a silent shore where a vast forest spread out over at least 700 or 800 square miles. In Stafford-shire, too, which is now almost the middle of England, another small coal-field tells the same story, while in South Wales the deep coal-mines and number of coal-seams remind us how for centuries and centuries forests must have flourished and have disappeared over and over again under the sand of the sea.

But what is it that has changed these beds of dead plants into hard, stony coal? In the first place you must remember they have been pressed down under an enormous weight of rocks above them. We can learn something about this even from our common lead pencils. At one time the graphite or pure carbon, of which the blacklead (as we wrongly call it) of our pencils is made, was dug solid out of the earth. but so much has now been used that they are obliged to collect the graphite dust, and press it under a heavy weight, and this makes such solid pieces that they can cut them into leads for ordinary cedar pencils.

Now the pressure which we can exert by machinery is absolutely nothing compared to the weight of all those hundreds of feet of solid rock which lie over the coal-beds, and which has pressed them down for thousands and perhaps millions of years; and besides this, we know that parts of the inside of the earth are very hot, and many of the rocks in which coal is found are altered by heat. So we can picture to ourselves that the coal was not only squeezed into a solid mass, but often much of the oil and gas which were in the leaves of the plants was driven out by heat, and the whole baked, as it were, into one substance. The difference between coal which flames and coal which burns only with a red heat, is chiefly that one has been baked and crushed more than the other. Coal which flames has still got in it the tar and the gas and the oils which the plant stored up in its leaves, and these when they escape again give back the sunbeams in a bright flame. The hard stone coal, on the contrary, has lost a great part of these oils, and only carbon remains, which seizes hold of the oxygen of the air and burns without flame. Coke is pure carbon, which we make artificially by driving out the oils and gases from coal, and the gas we burn is part of what is driven out.

We can easily make coal-gas here in this room. I have brought a tobacco-pipe, the bowl of which is filled with a little powdered coal, and the broad end cemented up with Stourbridge clay. When we place this bowl over a spirit-lamp and make it very hot, the gas is driven out at the narrow end of the pipe and lights easily (see Fig. 53). This is the way all our gas is made, only that furnaces are used to bake the coal in, and the gas is passed into large reservoirs till it is wanted for use.

You will find it difficult at first to understand how coal can be so full of oil and tar and gases, until you have tried to think over how much of all these there is in plants, and especially in seeds - think of the oils of almonds, of lavender, of cloves, and of caraways; and the oils of turpentine which we get from the pines, and out of which tar is made. When you remember these and many more, and also how the seeds of the club-moss now are largely charged with oil, you will easily imagine that the large masses of coal-plants which have been pressed together and broken and crushed, would give out a great deal of oil which, when made very hot, rises up as gas. You may often yourself see tar oozing out of the lumps of coal in a fire, and making little black bubbles which burst and burn. It is from this tar that James Young first made the paraffin oil we burn in our lamps, and the spirit benzoline comes from the same source.

From benzoline, again, we get a liquid called aniline, from which are made so many of our beautiful dyes - mauve, magenta, and violet; and what is still more curious, the bitter almonds, pear- drops, and many other sweets which children like to well, are actually flavoured by essences which come out of coal-tar. Thus from coal we get not only nearly all our heat and our light, but beautiful colours and pleasant

flavours. We spoke just now of the plants of the coal as being without beautiful flowers, and yet we see that long, long after their death they give us lovely colours and tints as beautiful as any in flower-world now.
Think, then, how much we owe to these plants which lived and died so long ago! If they had been able to reason, perhaps they might have said that they did not seem of much use in the world. They had no pretty flowers, and there was no one to admire their beautiful green foliage except a few croaking reptiles, and little crickets and grasshoppers; and they lived and died all on one spot, generation after generation, without seeming to do much good to anything or anybody. Then they were covered up and put out of sight, and down in the dark earth they were pressed all out of shape and lost their beauty and became only black, hard coal. There they lay for centuries and centuries, and thousands and thousands of years, and still no one seemed to want them.
At last, one day, long, long after man had been living on the earth, and had been burning wood for fires, and so gradually using up the trees in the forests, it was discovered that this black stone would burn, and from that time coal has been becoming every day more and more useful. Without it not only should we have been without warmth in our houses, or light in our streets when the stock of forest-wood was used up; but we could never have melted large quantities of iron-stone and extracted the iron. We have proof of this in Sussex. The whole country is full of iron-stone, and the railings of St. Paul's churchyard are made of Sussex iron. Iron-foundries were at work there as long as there was wood enough to supply them, but gradually the works fell into disuse, and the last furnace was put out in the year 1809. So now, because there is no coal in Sussex, the iron lies idle, while in the North, where the iron-stone is near the coal- mines, hundreds of tons are melted out every day.
Again, without coal we could have had no engines of any kind, and consequently no large manufactories of cotton goods, linen goods, or cutlery. In fact, almost everything we use could only have been made with difficulty and in small quantities; and even if we could have made them it would have been impossible to have sent them so quickly all over the world without coal, for we could have had no railways or steamships, but must have carried all goods along canals, and by slow sailing vessels. We ourselves must have taken days to perform journeys now made in a few hours, and months to reach our colonies.
In consequence of this we should have remained a very poor people. Without manufactories and industries we should have had to live chiefly by tilling the ground, and everyone being obliged to toil for daily bread, there would have been much less time or opportunity for anyone to study science, or literature, or history, or to provide themselves with comforts

and refinements of life.
All this then, those plants and trees of the far-off ages, which seemed to lead such useless lives, have done and are doing for us. There are many people in the world who complain that life is dull, that they do not see the use of it, and that there seems no work specially for them to do. I would advise such people, whether they are grown up or little children, to read the story of the plants which form the coal. These saw no results during their own short existences, they only lived and enjoyed the bright sunshine, and did their work, and were content. And now thousands, probably millions, of years after they lived and died, England owes her greatness, and we much of our happiness and comfort, to the sunbeams which those plants wove into their lives.
They burst forth again in our fires, in our brilliant lights, and in our engines, and do the greater part of our work; teaching us

"That nothing walks with aimless feet
That not one life shall be destroyed,
Or cast as rubbish to the void,
When God hath made the pile complete."

In Memoriam

Week 25
Lecture IX
Bees in the Hive

I am going to ask you to visit with me to-day one of the most wonderful cities with no human beings in it, and yet it is densely populated, for such a city may contain from twenty thousand to sixty thousand inhabitants. In it you will find streets, but no pavements, for the inhabitants walk along the walls of the houses; while in the houses you will see no windows, for each house just fits its owner, and the door is the only opening in it. Though made without hands these houses are most evenly and regularly built in tiers one above the other; and here and there a few royal palaces, larger and more spacious than the rest, catch the eye conspicuously as they stand out at the corners of the streets.
Some of the ordinary houses are used to live in, while others serve as storehouses where food is laid up in the summer to feed the inhabitants during the winter, when they are not allowed to go outside the walls. Not that the gates are ever shut: that is not necessary, for in this wonderful city each citizen follows the laws; going out when it is time to go out, coming home at proper hours, and staying at home when it is his or her duty. And in the winter, when it is very cold outside, the inhabitants,

having no fires, keep themselves warm within the city by clustering together, and never venturing out of doors.
One single queen reigns over the whole of this numerous population, and you might perhaps fancy that, having so many subjects to work for her and wait upon her, she would do nothing but amuse herself. On the contrary, she too obeys the laws laid down for her guidance, and never, except on one or two state occasions, goes out of the city, but works as hard as the rest in performing her own royal duties.
From sunrise to sunset, whenever the weather is fine, all is life, activity, and bustle in this busy city. Though the gates are so narrow that two inhabitants can only just pass each other on their way through them, yet thousands go in and out every hour of the day; some bringing in materials to build new houses, others food and provisions to store up for the winter; and while all appears confusion and disorder among this rapidly moving throng, yet in reality each has her own work to do, and perfect order reigns over the whole.
Even if you did not already know from the title of the lecture what city this is that I am describing, you would no doubt guess that it is a beehive. For where in the whole world, except indeed upon an anthill, can we find so busy, so industrious, or so orderly a community as among the bees? More than a hundred years ago, a blind naturalist, Francois Huber, set himself to study the habits of these wonderful insects and with the help of his wife and an intelligent manservant managed to learn most of their secrets. Before his time all naturalists had failed in watching bees, because if they put them in hives with glass windows, the bees, not liking the light, closed up the windows with cement before they began to work. But Huber invented a hive which he could open and close at will, putting a glass hive inside it, and by this means he was able to surprise the bees at their work. Thanks to his studies, and to those of other naturalists who have followed in his steps, we now know almost as much about the home of bees as we do about our own; and if we follow out to-day the building of a bee-city and the life of its inhabitants, I think you will acknowledge that they are a wonderful community, and that it is a great compliment to anyone to say that he or she is "as busy as a bee."
In order to begin at the beginning of the story, let us suppose that we go into a country garden one fine morning in May when the sun is shining brightly overhead, and that we see hanging from the bough of an old apple-tree a black object which looks very much like a large plum-pudding. On approaching it, however, we see that it is a large cluster or swarm of bees clinging to each other by their legs; each bee with its two fore-legs clinging to the two hinder legs of the one above it. In this way as many as 20,000 bees may be clinging together, and yet they hang so

freely that a bee, even from quite the centre of the swarm, can disengage herself from her neighbours and pass through to the outside of the cluster whenever she wishes.

If these bees were left to themselves, they would find a home after a time in a hollow tree, or under the roof of a house, or in some other cavity, and begin to build their honeycomb there. But as we do not wish to lose their honey we will bring a hive, and, holding it under the swarm, shake the bough gently so that the bees fall into it, and cling to the sides as we turn it over on a piece of clean linen, on the stand where the hive is to be.

And now let us suppose that we are able to watch what is going on in the hive. Before five minutes are over the industrious little insects have begun to disperse and to make arrangements in their new home. A number (perhaps about two thousand) of large, lumbering bees of a darker colour than the rest, will it is true, wander aimlessly about the hive, and wait for the others to feed them and house them; but these are the drones, or male bees (3, Fig. 54), who never do any work except during one or two days in their whole lives. But the smaller working bees (1, Fig. 54) begin to be busy at once. Some fly off in search of honey. Others walk carefully all round the inside of the hive to see if there are any cracks in it; and if there are, they go off to the horse-chestnut trees, poplars, hollyhocks, or other plants which have sticky buds, and gather a kind of gum called "propolis," with which they cement the cracks and make them air-tight. Others again, cluster round one bee (2, Fig. 54) blacker than the rest and having a longer body and shorter wings; for this is the queen-bee, the mother of the hive, and she must be watched and tended.

But the largest number begin to hang in a cluster from the roof just as they did from the bough of the apple tree. What are they doing there? Watch for a little while and you will soon see one bee come out from among its companions and settle on the top of the inside of the hive, turning herself round and round, so as to push the other bees back, and to make a space in which she can work. Then she will begin to pick at the under part of her body with her fore-legs, and will bring a scale of wax from a curious sort of pocket under her abdomen. Holding this wax in her claws, she will bite it with her hard, pointed upper jaws, which move to and fro sideways like a pair of pincers, then, moistening it with her tongue into a kind of paste, she will draw it out like a ribbon and plaster it on the top of the hive.

After that she will take another piece; for she has eight of these little wax-pockets, and she will go on till they are all exhausted. Then she will fly away out of the hive, leaving a small lump on the hive ceiling or on the bar stretched across it; then her place will be taken by another bee who will go through the same manoeuvres. This bee will be followed by another, and another, till a large wall of wax has been built, hanging from the bar of the hive as in Fig. 55, only that it will not yet have cells fashioned in it.

Meanwhile the bees which have been gathering honey out of doors begin to come back laden. But they cannot store their honey, for there are no cells made yet to put it in; neither can they build combs with the rest, for they have no wax in their wax-pockets. So they just go and hang quietly on to the other bees, and there they remain for twenty-four hours, during which time they digest the honey they have gathered, and part of it forms wax and oozes out from the scales under their body. Then they are prepared to join the others at work and plaster wax on to the hive.

Week 26

And now, as soon as a rough lump of wax is ready, another set of bees come to do their work. These are called the nursing bees, because they prepare the cells and feed the young ones. One of these bees, standing on the roof of the hive, begins to force her head into the wax, biting with her jaws and moving her head to and fro. Soon she has made the beginning of a round hollow, and then she passes on to make another, while a second bee takes her place and enlarges the first one. As many as twenty bees will be employed in this way, one after another, upon each hole before it is large enough for the base of a cell.

Meanwhile another set of nursing bees have been working just in the same way on the other side of the wax, and so a series of hollows are made back to back all over the comb. Then the bees form the walls of the cells and soon a number of six-sided tubes, about half an inch deep, stand all along each side of the comb ready to receive honey or bee-eggs.

You can see the shape of these cells in c,d, Fig. 56, and notice how closely they fit into each other. Even the ends are so shaped that, as they lie back to back, the bottom of one cell (B, Fig. 56) fits into the space between the ends of three cells meeting it from the opposite side (A, Fig. 56), while they fit into the spaces around it. Upon this plan the clever little bees fill every atom of space, use the least possible quantity of wax, and make the cells lie so closely together that the whole comb is kept warm when the young bees are in it.

There are some kinds of bees who do not live in hives, but each one builds a home of its own. These bees - such as the upholsterer bee, which digs a hole in the earth and lines it with flowers and leaves, and the mason bee, which builds in walls - do not make six-sided cells, but round ones, for room is no object to them. But nature has gradually taught the little hive-bee to build its cells more and more closely, till they fit perfectly within each other. If you make a number of round holes close together in a soft substance, and then squeeze the substance evenly from all sides, the rounds will gradually take a six-sided form, showing that this is the closest shape into which they can be compressed. Although the bee does not know this, yet as gnaws away every bit of wax that can be spared she brings the holes into this shape.

As soon as one comb is finished, the bees begin another by the side of it, leaving a narrow lane between, just broad enough for two bees to pass back to back as they crawl along, and so the work goes on till the hive is full of combs.

As soon, however, as a length of about five or six inches of the first comb has been made into cells, the bees which are bringing home honey no longer hang to make it into wax, but begin to store it in the cells. We all know where the bees go to fetch their honey, and how, when a bee settles on a flower, she thrusts into it her small tongue-like proboscis, which is really a lengthened under-lip, and sucks out the drop of honey. This she swallows, passing it down her throat into a honey-bag or first stomach, which lies between her throat and her real stomach, and when she gets back to the hive she can empty this bag and pass honey back through her mouth again into the honey-cells.

But if you watch bees carefully, especially in the spring-time, you will find that they carry off something else besides honey. Early in the morning, when the dew is on the ground, or later in the day, in moist shady places, you may see a bee rubbing itself against a flower, or biting those bags of yellow dust or pollen which we mentioned in Lecture VII. When she has covered herself with pollen, she will brush it off with her feet, and, bringing it to her mouth, she will moisten and roll it into a little ball, and then pass it back from the first pair of legs to the second and so to the third or hinder pair. Here she will pack it into a little hairy groove called a "basket" in the joint of one of the hind legs, where you may see it, looking like a swelled joint, as she hovers among the flowers. She often fills both hind legs in this way, and when she arrives back at the hive the nursing bees take the lumps form her, and eat it themselves, or mix it with honey to feed the young bees; or, when they have any to spare, store it away in old honey-cells to be used by-and-by. This is the dark, bitter stuff called "bee- bread" which you often find in a honeycomb, especially in a comb which has been filled late in the summer.

When the bee has been relieved of the bee-bread she goes off to one of the clean cells in the new comb, and, standing on the edge, throws up the honey from the honey-bag into the cell. One cell will hold the contents of many honey-bags, and so the busy little workers have to work all day filling cell after cell, in which the honey lies uncovered, being too thick and sticky to flow out, and is used for daily food - unless there is any to spare, and then they close up the cells with wax to keep for the winter.

Meanwhile, a day or two after the bees have settled in the hive, the queen-bee begins to get very restless. She goes outside the hive and hovers about a little while, and then comes in again, and though generally the bees all look very closely after her to keep her indoors, yet now they let her do as she likes. Again she goes out, and again back, and then, at last, she soars up into the air and flies away. But she is not allowed to go alone. All the drones of the hive rise up after her, forming a guard of honour to follow her wherever she goes.

In about half-an-hour she comes back again, and then the working bees all gather round her, knowing that now she will remain quietly in the hive and spend all her time in laying eggs; for it is the queen-bee who lays all the eggs in the hive. This she begins to do about two days after her flight. There are now many cells ready besides those filled with honey; and, escorted by several bees, the queen-bee goes to one of these, and, putting her head into it remains there a second as if she were examining whether it would make a good home for the young bee. Then, coming out, she turns round and lays a small, oval, bluish-white egg in the cell. After this she takes no more notice of it, but goes on to the next cell and the next, doing the same thing, and laying eggs in all the empty cells equally on both sides of the comb. She goes on so quickly that she sometimes lays as many as 200 eggs in one day.

Then the work of the nursing bees begins. In two or three days each egg has become a tiny maggot or larva, and the nursing bees put into its cell a mixture of pollen and honey which they have prepared in their own mouths, thus making a kind of sweet bath in which the larva lies. In five or six days the larva grows so fat upon this that it nearly fills the cell, and then the bees seal up the mouth of the cell with a thin cover of wax, made of little rings and with a tiny hole in the centre.

As soon as the larva is covered in, it begins to give out from its under-lip a whitish, silken film, made of two threads of silk glued together, and with this it spins a covering or cocoon all round itself, and so it remains for about ten days more. At last, just twenty-one days after the egg was laid, the young bee is quite perfect, lying in the cell as in Fig. 57, and she begins to eat her way through the cocoon and through the waxen lid, and scrambles out of her cell. Then the nurses come again to her, stroke her wings and feed her for twenty-four hours, and after that she is quite ready to begin work, and flies out to gather honey and pollen like the rest of the workers.

By this time the number of working bees in the hive is becoming very great, and the storing of honey and pollen-dust goes on very quickly. Even the empty cells which the young bees have left are cleaned out by the nurses and filled with honey; and this honey is darker than that stored in clean cells, and which we always call "virgin honey" because it is so pure and clear.

At last, after six weeks, the queen leaves off laying worker- eggs, and begins to lay, in some rather larger cells, eggs from which drones, or male bees, will grow up in about twenty days. Meanwhile the worker-bees have been building on the edge of the cones some very curious cells (q, Fig. 57) which look like thimbles hanging with the open side upwards, and about every three days the queen stops in laying drone-eggs and goes to put an egg in one of these cells. Notice that she waits three days between each of these peculiar layings, because we shall see presently that there is a good reason for her doing so.

The nursing bees take great care of these eggs, and instead of putting ordinary food into the cell, they fill it with a sweet, pungent jelly, for this larva is to become a princess and a future queen bee. Curiously enough, it seems to be the peculiar food and the size of the cell which makes the larva grow into a mother-bee which can lay eggs, for if a hive has the misfortune to lose its queen, they take one of the ordinary worker-larvae and put it into a royal cell and feed it with jelly, and it becomes a queen-bee. As soon as the princess is shut in like the others, she begins to spin her cocoon, but she does not quite close it as the other bees do, but leaves a hole at the top.

Week 27

At the end of sixteen days after the first royal egg was laid, the eldest princess begins to try to eat her way out of her cell, and about this time the old queen becomes very uneasy, and wanders about distractedly. The reason of this is that there can never be two queen-bees in one hive, and the queen knows that her daughter will soon be coming out of her cradle and will try to turn her off her throne. So, not wishing to have to fight for her kingdom, she makes up her mind to seek a new home and take a number of her subjects with her. If you watch the hive about this time you will notice many of the bees clustering together after they have brought in their honey, and hanging patiently, in order to have plenty of wax ready to use when they start, while the queen keeps a sharp look-out for a bright, sunny day, on which they can swarm: for bees will never swarm on a wet or doubtful day if they can possibly help it, and we can easily understand why, when we consider how the rain would clog their wings and spoil the wax under their bodies.

Meanwhile the young princess grows very impatient, and tries to get out of her cell, but the worker-bees drive her back, for they know there would be a terrible fight if the two queens met. So they close up the hole she has made with fresh wax after having put in some food for her to live upon till she is released.

At last a suitable day arrives, and about ten or eleven o'clock in the morning the old queen leaves the hive, taking with her about 2000 drones and from 12,000 to 20,000 worker-bees, which fly a little way clustering round her till she alights on the bough of some tree, and then they form a compact swarm ready for a new hive or to find a home of their own.

Leaving them to go their way, we will now return to the old hive. Here the liberated princess is reigning in all her glory; the worker-bees crowd round her, watch over her, and feed her as though they could not do enough to show her honour. But still she is not happy. She is restless, and runs about as if looking for an enemy, and she tries to get at the remaining royal cells where the other young princesses are still shut in. But the workers will not let her touch them, and at last she stands still and begins to beat the air with her wings and to tremble all over, moving more and more quickly, till she makes quite a loud, piping noise.

Hark! What is that note answering her? It is a low, hoarse sound, and it comes from the cell of the next eldest princess. Now we see why the young queen has been so restless. She knows her sister will soon come out, and the louder and stronger the sound becomes within the cell, the sooner she knows the fight will have to begin. And so she makes up her mind to follow her mother's example and to lead off a second swarm. But she cannot always stop to choose a fine day, for her sister is growing very strong and may come out of her cell before she is off. And so the second, or after swarm, gets ready and goes away. And this explains why princesses' eggs are laid a few days apart, for if they were laid all on the same day, there would be no time for one princess to go off with a swarm before the other came out of her cell. Sometimes, when the workers are not watchful enough, two queens do meet, and then they fight till one is killed; or sometimes they both go off with the same swarm without finding each other out. But this only delays the fight till they get into the new hive; sooner or later one must be killed.

And now a third queen begins to reign in the old hive, and she is just as restless as the preceding ones, for there are still more princesses to be born. But this time, if no new swarm wants to start, the workers do not try to protect the royal cells. The young queen darts at the first she sees, gnaws a hole with her jaws, and, thrusting in her sting through the hole in the cocoon, kills the young bee while it is still a prisoner. She then goes to the next, and the next, and never rests till all the young princesses are destroyed. Then she is contented, for she knows no other queen will come to dethrone her. After a few days she takes her flight in the air with the drones, and comes home to settle down in the hive for the winter.

Then a very curious scene takes place. The drones are no more use, for the queen will not fly out again, and these idle bees will never do any work in the hive. So the worker-bees begin to kill them, falling upon them, and stinging them to death, and as the drones have no stings they cannot defend themselves, and in a few days there is not a drone, nor even a drone-egg, left in the hive. This massacre seems very sad to us, since the poor drones have never done any harm beyond being hopelessly idle. But it is less sad when we know that they could not live many weeks, even if they were not attacked, and, with winter coming, the bees cannot afford to feed useless mouths, so a quick death is probably happier for them than starvation.

And now all the remaining inhabitants of the hive settle down to feeding the young bees and laying in the winter's store. It is at this time, after they have been toiling and saving, that we come and take their honey; and from a well-stocked hive we may even take 30 lbs. without starving the industrious little inhabitants. But then we must often feed them in return and give them sweet syrup in the late autumn and the next early spring when they cannot find any flowers.

Although the hive has now become comparatively quiet and the work goes on without excitement, yet every single bee is employed in some way, either out of doors or about the hive. Besides the honey collectors and the nurses, a certain number of bees are told off to ventilate the hive. You will easily understand that where so many insects are packed closely together the heat will become very great, and the air impure and unwholesome. And the bees have no windows that they can open to let in fresh air, so they are obliged to fan it in from the one opening of the hive. The way in which they do this is very interesting. Some of the bees stand close to the entrance, with their faces towards it, and opening their wings, so as to make them into fans, they wave them to and fro, producing a current of air. Behind these bees, and all over the floor of the hive, there stand others, this time with their backs towards the entrance, and fan in the same manner, and in this way air is sent into all the passages.

Another set of bees clean out the cells after the young bees are born, and make them fit to receive honey, while others guard the entrance of the hive to keep away the destructive wax-moth, which tries to lay its eggs in the comb so that its young ones may feed on the honey. All industrious people have to guard their property against thieves and vagabonds, and the bees have many intruders, such as wasps and snails and slugs, which creep in whenever they get a chance. If they succeed in escaping the sentinel bees, then a fight takes place within the hive, and the invader is stung to death.

Sometimes, however, after they have killed the enemy, the bees cannot get rid of his body, for a snail or slug is too heavy to be easily moved, and yet it would make the hive very unhealthy to allow it to remain. In this dilemma the ingenious little bees fetch the gummy "propolis" from the plant-buds and cement the intruder all over, thus embalming his body and preventing it from decaying.

And so the life of this wonderful city goes on. Building, harvesting, storing, nursing, ventilating and cleaning from morn till night, the little worker-bee lives for about eight months, and in that time has done quite her share of work in the world. Only the young bees, born late in the season, live on till the next year to work in the spring. The queen-bee lives longer, probably about two years, and then she too dies, after having had a family of many thousands of children.

We have already pointed out that in our fairy-land of nature all things work together so as to bring order out of apparent confusion. But though we should naturally expect winds and currents, rivers and clouds, and even plants to follow fixed laws, we should scarcely have looked for such regularity in the life of the active, independent busy bee. Yet we see that she, too, has her own appointed work to do, and does it regularly and in an orderly manner. In this lecture we have been speaking entirely of the bee within the hive, and noticing how marvellously her instincts guide her in her daily life. But within the last few years we have learnt that she performs a most curious and wonderful work in the world outside her home and that we owe to her not only the sweet honey to eat, but even in a great degree the beauty and gay colours of the flowers which she visits when collecting it. This work will form the subject of our next lecture, and while we love the little bee for her constant industry, patience, and order within the hive, we shall, I think, marvel at the wonderful law of nature which guides her in her unconscious mission of love among the flowers which grow around it.

Week 28

Lecture X

BEES AND FLOWERS

Whatever thoughts each one of you may have brought to the lecture to-day, I want you to throw them all aside and fancy yourself to be in a pretty country garden on a hot summer's morning. Perhaps you have been walking, or reading, or playing, but it is getting too hot now to do anything; and so you have chosen the shadiest nook under the old walnut-tree, close to the flower-bed on the lawn, and would almost like to go to sleep if it were not too early in the day.

As you lie there thinking of nothing in particular, except how pleasant it is to be idle now and then, you notice a gentle buzzing close to you, and you see that on the flower-bed close by, several bees are working busily among the flowers. They do not seem to mind the heat, nor to wish to rest; and they fly so lightly and look so happy over their work that it does not tire you to look at them.

That great humble-bee takes it leisurely enough as she goes lumbering along, poking her head into the larkspurs, and remaining so long in each you might almost think she had fallen asleep. The brown hive-bee on the other hand, moves busily and quickly among the stocks, sweet peas, and mignonette. She is evidently out on active duty, and means to get all she can from each flower, so as to carry a good load back to the hive. In some blossoms she does not stay a moment, but draws her head back directly she has popped it in, as if to say "No honey there." But over the full blossoms she lingers a little, and then scrambles out again with her drop of honey, and goes off to seek more in the next flower.

Let us watch her a little more closely. There are plenty of different plants growing in the flower-bed, but, curiously enough, she does not go first to one kind and then to another; but keeps to one, perhaps the mignonette, the whole time till she flies away. Rouse yourself up to follow her, and you will see she takes her way back to the hive. She may perhaps stop to visit a stray plant of mignonette on her way, but no other flower will tempt her till she has taken her load home.

Then when she comes back again she may perhaps go to another kind of flower, such as the sweet peas, for instance, and keep to them during the next journey, but it is more likely that she will be true to her old friend the mignonette for the whole day.

We all know why she makes so many journeys between the garden and the hive, and that she is collecting drops of honey from each flower, and carrying it to be stored up in the honeycomb for winter's food. How she stores it, and how she also gathers pollen-dust for her bee-bread, we saw in the last lecture; to-day we will follow her in her work among the flowers, and see, while they are so useful to her, what she is doing for them in return.

We have already learnt from the life of a primrose that plants can make better and stronger seeds when they can get pollen-dust from another plant, than when they are obliged to use that which grows in the same flower; but I am sure you will be very much surprised to hear that the more we study flowers the more we find that their colours, their scent, and their curious shapes are all so many baits and traps set by nature to entice insects to come to the flowers, and carry this pollen-dust from one to the other.

So far as we know, it is entirely for this purpose that the plants form honey in different parts of the flower, sometimes in little bags or glands, as in the petals of the buttercup flower, sometimes in clear drops, as in the tube of the honeysuckle. This food they prepare for the insects, and then they have all sorts of contrivances to entice them to come and fetch it.

You will remember that the plants of the coal had no bright or conspicuous flowers. Now we can understand why this was, for there were no flying insects at that time to carry the pollen- dust from flower to flower, and therefore there was no need of coloured flowers to attract them. But little by little, as flies, butterflies, moths and bees began to live in the world, flowers too began to appear, and plants hung out these gay-coloured signs, as much as to say, "Come to me, and I will give you honey if you will bring me pollen-dust in exchange, so that my seeds may grow healthy and strong."

We cannot stop to inquire to-day how this all gradually came about, and how the flowers gradually put on gay colours and curious shapes to tempt the insects to visit them; but we will learn something about the way they attract them now, and how you may see it for yourselves if you keep your eyes open.

For example, if you watch the different kinds of grasses, sedges and rushes, which have such tiny flowers that you can scarcely see them, you will find that no insects visit them. Neither will you ever find bees buzzing round oak-trees, nut-trees, willows, elms or birches. But on the pretty and sweet-smelling apple- blossoms, or the strongly scented lime-trees, you will find bees, wasps, and plenty of other insects.

The reason of this is that grasses, sedges, rushes, nut-trees, willow, and the others we have mentioned, have all of them a great deal of pollen-dust, and as the wind blows them to and fro, it wafts the dust from one flower to another, and so these plants do not want the insects, and it is not worth their while to give out honey, or to have gaudy or sweet-scented flowers to attract them.

But wherever you see bright or conspicuous flowers you may be quite sure that the plants want the bees or some other winged insect to come and carry their pollen for them. Snowdrops hanging their white heads among their green leaves, crocuses with their violet and yellow flowers, the gaudy poppy, the large- flowered hollyhock or the sunflower, the flaunting dandelion, the pretty pink willow-herb, the clustered blossoms of the mustard and turnip flowers, the bright blue forget-me-not and the delicate little yellow trefoil, all these are visited by insects, which easily catch sight of them as they pass by and hasten to sip their honey.

Sir John Lubbock has shown that bees are not only attracted by bright colours, but that they even know one colour from another. He put some honey on slips of glass with coloured papers under them, and when he had accustomed the bees to find the honey always on the blue glass, he washed this glass clean, and put the honey on the red glass instead. Now if the bees had followed only the smell of the honey, they would have flown to the red glass, but they did not. They went first to the blue glass, expecting to find the honey on the usual colour, and it was only when they were disappointed that they went off to the red.

Is it not beautiful to think that the bright pleasant colours we love so much in flowers, are not only ornamental, but that they are useful and doing their part in keeping up healthy life in our world?

Neither must we forget what sweet scents can do. Have you never noticed the delicious smell which comes from beds of mignonette, thyme, rosemary, mint, or sweet alyssum, from the small hidden bunches of laurustinus blossom, or from the tiny flowers of the privet? These plants have found another way of attracting the insects; they have no need of bright colours, for their scent is quite as true and certain a guide. You will be surprised if you once begin to count them up, how many white and dull or dark- looking flowers are sweet-scented, while gaudy flowers, such as tulip, foxglove and hollyhock, have little or no scent. And then, just as in the world we find some people who have everything to attract others to them, beauty and gentleness, cleverness, kindliness, and loving sympathy, so we find some flowers, like the beautiful lily, the lovely rose, and the delicate hyacinth, which have colour and scent and graceful shapes all combined.

But we are not yet nearly at an end of the contrivances of flowers to secure the visits of insects. Have you not observed that different flowers open and close at different times? The daisy receives its name day's eye, because it opens at sunrise and closes at sunset, while the evening primrose (Aenothera biennis) and the night campion (Silene noctiflora) spread out their flowers just as the daisy is going to bed.

What do you think is the reason of this? If you go near a bed of evening primroses just when the sun is setting, you will soon be able to guess, for they will then give out such a sweet scent that you will not doubt for a moment that they are calling the evening moths to come and visit them. The daisy opens by day, because it is visited by day insects, but those particular moths which can carry the pollen-dust of the evening primrose, fly only by night, and if this flower opened by day other insects might steal its honey, while they would not be the right size or shape to touch its pollen-bags and carry the dust.

It is the same if you pass by a honeysuckle in the evening; you will be surprised how much stronger its scent is than in the day- time. This is because the sphinx hawk-moth is the favourite visitor of that flower, and comes at nightfall, guided by the strong scent, to suck out the honey with its long proboscis, and carry the pollen-dust.

Again, some flowers close whenever rain is coming. The pimpernel (Anagallis arvensis) is one of these, hence its name of the "Shepherd's Weather-glass." This little flower closes, no doubt, to prevent its pollen-dust being washed away, for it has no honey; while other flowers do it to protect the drop of honey at the bottom of their corolla. Look at the daisies for example when a storm is coming on; as the sky grows dark and heavy, you will see them shrink up and close till the sun shines again. They do this because in each of the little yellow florets in the centre of the flower there is a drop of honey which would be quite spoiled if it were washed by the rain.

And now you will see why cup-shaped flowers so often droop their heads - think of the harebell, the snowdrop, the lily-of-the- valley, the campanula, and a host of others; how pretty they look with their bells hanging so modestly from the slender stalk! They are bending down to protect the honey-glands within them, for if the cup became full of rain or dew the honey would be useless, and the insects would cease to visit them.

Week 29

But it is not only necessary that the flowers should keep their honey for the insects, they also have to take care and keep it for the right kind of insect. Ants are in many cases great enemies to them, for they like honey as much as bees and butterflies do, yet you will easily see that they are so small that if they creep into a flower they pass the anthers without rubbing against them, and so take the honey without doing any good to the plant. Therefore we find numberless contrivances for keeping the ants and other creeping insects away. Look for example at the hairy stalk of the primrose flower; those little hairs are like a forest to a tiny ant, and they protect the flower from his visits. The Spanish catchfly (Silene otites), on the other hand, has a smooth, but very gummy stem, and on this the insects stick, if they try to climb. Slugs and snails too will often attack and bite flowers, unless they are kept away by thorns and bristles, such as we find on the teazel and the burdock. And so we are gradually learning that everything which a plant does has its meaning, if we can only find it out, and that even very insignificant hair has its own proper use, and when we are once aware of this a flower-garden may become quite a new world to us if we open our eyes to all that is going on in it.

But as we cannot wander among many plants to-day, let us take a few which the bees visit, and see how they contrive not to give up their honey without getting help in return. We will start with the blue wood-geranium, because from it we first began to learn the use of insects to flowers.

More than a hundred years ago a young German botanist, Christian Conrad Sprengel, noticed some soft hairs growing in the centre of this flower, just round the stamens, and he was so sure that every part of a plant is useful, that he set himself to find out what these hairs meant. He soon discovered that they protected some small honey-bags at the base of the stamens, and kept the rain from washing the honey away, just as our eyebrows prevent the perspiration on our faces from running into our eyes. This led him to notice that plants take great care to keep their honey for insects, and by degrees he proved that they did this in order to tempt the insects to visit them and carry off their pollen.

The first thing to notice in this little geranium flower is that the purple lines which ornament it all point directly to the place where the honey lies at the bottom of the stamens, and actually serve to lead the bee to the honey; and this is true of the veins and marking of nearly all flowers except of those which open by night, and in these they would be useless, for the insects would not see them.

When the geranium first opens, all its ten stamens are lying flat on the corolla or coloured crown, as in the left-hand flower in Fig. 58, and then the bee cannot get at the honey. But in a short time five stamens begin to raise themselves and cling round the stigma or knob at the top of the seed-vessel, as in the middle flower. Now you would think they would leave their dust there. But no! the stigma is closed up so tight that the dust cannot get on to the sticky part. Now, however, the bee can get at the honey-glands on the outside of the raised stamens; and as he sucks it, his back touches the anthers or dust-bags, and he carries off the pollen. Then, as soon as all their dust is gone, these five stamens fall down, and the other five spring up. Still, however, the stigma remains closed, and the pollen of these stamens, too, may be carried away to another flower. At last these five also fall down, and then, and not till then, the stigma opens and lays out its five sticky points, as you may see in the right-hand flower, Fig. 58.

But its own pollen is all gone, how then will it get any? It will get it from some bee who has just taken it from another and younger flower; and thus you see the blossom is prevented from using its own pollen, and made to use that of another blossom, so that its seeds may grow healthy and strong.

The garden nasturtium, into whose blossom we saw the humble-bee poling his head, takes still more care of its pollen-dust. It hides its honey down at the end of its long spur, and only sends out one stamen at a time instead of five like the geranium; and then, when all the stamens have had their turn, the sticky knob comes out last for pollen from another flower.

All this you may see for yourselves if you find geraniums* in the hedges, and nasturtiums in you garden. But even if you have not these, you may learn the history of another flower quite as curious, and which you can find in any field or lane even near London. The common dead-nettle (Fig. 59) takes a great deal of trouble in order that the bee may carry off its pollen. When you have found one of these plants, take a flower from the ring all round the stalk and tear it gently open, so that you can see down its throat. There, just at the very bottom, you will find a thick fringe of hairs, and you will guess at once that these are to protect a drop of honey below. Little insects which would creep into the flower and rob it of its honey without touching the anthers of the stamens cannot get past these hairs, and so the drop is kept till the bee comes to fetch it. (*The scarlet and other bright geraniums of our flower-gardens are not true geraniums, but pelargoniums. You may, however, watch all these peculiarities in them if you cannot procure the true wild geranium.)

Now look for the stamens; there are four of them, two long and two short, and they are quite hidden under the hood which forms the top of the flower. How will the bee touch them? If you were to watch one, you would find that when the bee alights on the broad lip and thrusts her head down the tube, she first of all knows her back against the little forked tip. This is the sticky stigma, and she leaves there any dust she has brought from another flower; then, as she must push far in to reach the honey, before she comes out again has carried away the yellow powder on her back, ready to give it to the next flower.

Do you remember how we noticed at the beginning of the lecture that a bee always likes to visit the same kind of plant in one journey? You see now that this is very useful to the flowers. If the bee went from a dead-nettle to a geranium, the dust would be lost, for it would be of no use to any other plant but a dead- nettle. But since the bee likes to get the same kind of honey each journey, she goes to the same kind of flowers, and places the pollen-dust just where it is wanted.

There is another flower, called the Salvia, which belongs to the same family as our dead-nettle, and I think you will agree with me that its way of dusting the bee's back is most clever. The Salvia (Fig. 60) is shaped just like the dead-nettle, with a hood and a broad lip, but instead of four stamens it has only two, the other two being shrivelled up. The two that are left have a very strange shape, for the stalk or filament of the stamen is very short, while the anther, which is in most flowers two little bags stuck together, has here grown out into a long thread, with a little dust-bag at one end only. In 1, Fig. 60, you only see one of these stems, because the flower is cut in half, but in the whole flower, one stands on each side just within the lip. Now, when the bee puts her head into the tube to reach the honey, she passes right between these two swinging anthers, and knocking against the end pushes it before her and so brings the dust-bag plump down on her back, scattering the dust there! you can easily try this by thrusting a pencil into any Salvia flower, and you will see the anther fall.

You will notice that all this time the be does not touch the sticky stigma which hangs high above her, but after the anthers are empty and shrivelled the stalk of the stigma grows longer, and it falls lower down. By-and-by another bee, having pollen on her back, comes to look for honey, and as she goes into No. 3, she rubs against the stigma and leaves upon it the dust from another flower.

Tell me, has not the Salvia, while remaining so much the same shape as the dead-nettle, devised a wonderful contrivance to make use of the visits of the bee?

The common sweet violet (Viola odorata) or the dog violet (Viola canina), which you can gather in any meadow, give up their pollen-dust in quite a different way from the Salvia, and yet it is equally ingenious. Everyone has noticed what an irregular shape this flower has, and that one of its purple petals has a curious spur sticking out behind. In the tip of this spur and in the spur of the stamen lying in it the violet hides its honey, and to reach it the bee must press past the curious ring of orange-tipped bodies in the middle of the flower. These bodies are the anthers, Fig. 61, which fit tightly round the stigma, so that when the pollen-dust, which is very dry, comes out of the bags, it remains shut in by the tips as if in a box. Two of these stamens have spurs which lie in the coloured spur of the flower, and have honey at the end of them. Now, when the bee shakes the end of the stigma, it parts the ring of anthers, and the fine dust falls through upon the insect.

Let us see for a moment how wonderfully this flower is arranged to bring about the carrying of the pollen, as Sprengel pointed out years ago. In the first place, it hangs on a thin stalk, and bends its head down so that the rain cannot come near the honey in the spur, and also so that the pollen-dust falls forward into the front of the little box made by the closed anthers. Then the pollen is quite dry, instead of being sticky as in most plants. This is in order that it may fall easily through the cracks. Then the style or stalk of the stigma is very thin and its tip very broad, so that it quivers easily when the bee touches it, and so shakes the anthers apart, while the anthers themselves fold over to make the box, and yet not so tightly but that the dust can fall through when they are shaken. Lastly, if you look at the veins of the flower, you will find that they all point towards the spur where the honey is to be found, so that when the sweet smell of the flower has brought the bee, she cannot fail to go in at the right place.

Two more flowers still I want us to examine together, and then I hope you will care to look at every flower you meet, to try and see what insects visit it, and how its pollen-dust is carried. These two flowers are the common Bird's-foot trefoil (Lotus corniculatus), and the Early Orchis (Orchis mascula), which you may find in almost any moist meadow in the spring and early summer.

The Bird's-foot trefoil, Fig. 62, you will find almost anywhere all through the summer, and you will know it from other flowers very like it by its leaf, which is not a true trefoil, for behind the three usual leaflets of the clover and the shamrock leaf, it has two small leaflets near the stalk. The flower, you will notice, is shaped very like the flower of a pea, and indeed it belongs to the same family, called the Papilionaceae or butterfly family, because the flowers look something like an insect flying. In all these flowers the top petal stands up like a flag to catch the eye of the insect, and for this reason botanists call it the "standard". Below it are two side-petals called the "wings," and if you pick these off you will find that the remaining two petals are joined together at the tip in a shape like the keel of a boat. For this reason they are called the "keel". Notice as we pass that these two last petals have in them a curious little hollow or depression, and if you look inside the "wings" you will notice a little knob that fits into this hollow, and so locks the two together. We shall see by-and-by that this is important.

Week 30

Next let us look at the half-flower when it is cut open, and see what there is inside. There are ten stamens in all, enclosed with the stigma in the keel; nine are joined together and one is by itself. The anthers of five of these stamens burst open while the flower is still a bud, but the other stamens go on growing, and push the pollen-dust, which is very moist and sticky, right up into the tip of the keel. Here you see it lies right round the stigma, but as we saw before in the geranium, the stigma is not ripe and sticky yet, and so it does not use the pollen grains.

Now suppose that a bee comes to the flower. The honey she has to fetch lies inside the tube, and the one stamen being loose she is able to get her proboscis in. but if she is to be of any use to the flower she must uncover the pollen-dust. See how cunningly the flower has contrived this. In order to put her head into the tube the bee must stand upon the wings, and her weight bends them down. but they are locked to the keel by the knob fitting in the hole, and so the keel is pushed down too, and the sticky pollen- dust is uncovered and comes right against the stomach of the bee and sticks there! As soon as she has done feeding and flies away, up go the wings and the keel with them, covering up any pollen that remains ready for next time. Then when the bee goes to another flower, as she touches the stigma as well as the pollen, she leaves some of the foreign dust upon it, and the flower uses that rather than its own, because it is better for its seeds. If however no bee happens to come to one of these flowers, after a time the stigma becomes sticky and it uses its own pollen: and this is perhaps one reason why the bird's-foot trefoil is so very common, because it can do its own work if the bee does not help it.

Now we come lastly to the Orchis flower. Mr. Darwin has written a whole book on the many curious and wonderful ways in which orchids tempt bees and other insects to fertilize them. We can only take the simplest, but I think you will say that even this blossom is more like a conjuror's box than you would have supposed it possible that a flower could be.

Let us examine it closely. It has sic deep-red covering leaves, Fig. 62, three belonging to the calyx or outer cup, and three belonging to the corolla or crown of the flower; but all six are coloured alike, except that the large on in front, called the "lip", has spots and lines upon it which will suggest to you at once that they point to the honey.

But where are the anthers, and where is the stigma? Look just under the arch made by those three bending flower-leaves, and there you will see two small slits, and in these some little club-shaped bodies, which you can pick out with the point of a needle. One of these enlarged is shown. It is composed of sticky grains of pollen held together by fine threads on the top of a thin stalk; and at the bottom of the stalk there is a little round body. This is all that you will find to represent the stamens of the flower. When these masses of pollen, or pollinia as they are called, are within the flower, the knob at the bottom is covered by a little lid, shutting them in like the lid of a box, and just below this lid you will see two yellowish lumps, which are very sticky. These are the top of the stigma, and they are just above the seed-vessel, which you can see in the lowest flower in the picture.

Now let us see how this flower gives up its pollen. When a bee comes to look for honey in the orchis, she alights on the lip, and guided by the lines makes straight for the opening just in front of the stigmas. Putting her head into this opening she pushes down into the spur, where by biting the inside skin she gets some juicy sap. Notice that she has to bite, which takes time.

You will see at once that she must touch the stigmas in going in, and so give them any pollen she has on her head. but she also touches the little lid and it flies instantly open, bringing the glands at the end of the pollen-masses against her head. These glands are moist and sticky, and while she is gnawing the inside of the spur they dry a little and cling to her head and she brings them out with her. Darwin once caught a bee with as many as sixteen of these pollen-masses clinging to her head.

But if the bee went into the next flower with these pollinia sticking upright, she would simply put them into the same slits in the next flower, she would not touch them against the stigma. Nature, however, has provided against this. As the bee flies along, the glands sticking to its head dry more and more, and as they dry they curl up and drag the pollen-masses down, so that instead of standing upright, as in 1, Fig. 63, they point forwards, as in 2.

And now, when the bee goes into the next flower, she will thrust them right against the sticky stigmas, and as they cling there the fine threads which hold the grains together break away, and the flower is fertilized.

If you will gather some of these orchids during your next spring walk in the woods, and will put a pencil down the tube to represent the head of the bee you may see the little box open, and the two pollen-masses cling to the pencil. Then if you draw it out you may see them gradually bend forwards, and by thrusting your pencil into the next flower you may see the grains of pollen bread away, and you will have followed out the work of a bee.

Do not such wonderful contrivances as these make us long to know and understand all the hidden work that is going on around us among the flowers, the insects, and all forms of life? I have been able to tell you but very little, but I can promise you that the more you examine, the more you will find marvellous histories such as these in simple field-flowers.

Long as we have known how useful honey was to the bee, and how it could only get it from flowers, yet it was not till quite lately that we have learned to follow out Sprengel's suggestion, and to trace the use which the bee is to the flower. But now that we have once had our eyes opened, every flower teaches us something new, and we find that each plant adapts itself in a most wonderful way to the insects which visit it, both so as to provide them with honey, and at the same time to make them unconsciously do it good service.

And so we learn that even among insects and flowers, those who do most for others, receive most in return. The bee and the flower do not either of them reason about the matter, they only go on living their little lives as nature guides them, helping and improving each other. Think for a moment how it would be, if a plant used up all its sap for its own life, and did not give up any to make the drop of honey in its flower. The bees would soon find out that these particular flowers were not worth visiting, and the flower would not get its pollen-dust carried, and would have to do its own work and grow weakly and small. Or suppose on the other hand that the bee bit a hole in the bottom of the flower, and so got at the honey, as indeed they sometimes do; then she would not carry the pollen-dust, and so would not keep up the healthy strong flowers which make her daily food.

But this, as you see, is not the rule. On the contrary, the flower feeds the bee, and the bee quite unconsciously helps the flower to make its healthy seed. Nay more; when you are able to read all that has been written on this subject, you will find that we have good reason to think that the flowerless plants of the Coal Period have gradually put on the beautiful colours, sweet scent, and graceful shapes of our present flowers, in consequence of the necessity of attracting insects, and thus we owe our lovely flowers to the mutual kindliness of plants and insects.

And is there nothing beyond this? Surely there is. Flowers and insects, as we have seen, act without thought or knowledge of what they are doing; but the law of mutual help which guides them is the same which bids you and me be kind and good to all those around us, if we would lead useful and happy lives. And when we see that the Great Power which rules over our universe makes each work for the good of all, even in such humble things as bees and flowers; and that beauty and loveliness come out of the struggle and striving of all living things; then, if our own life be sometimes difficult, and the struggle hard to bear, we learn from the flowers that the best way to meet our troubles is to lay up our little drop of honey for others, sure that when they come to sip it they will, even if unconsciously, give us new vigour and courage in return.

And now we have arrived at the end of those subjects which we selected out of the Fairy-land of Science. You must not for a moment imagine, however, that we have in any way exhausted our fairy domain; on the contrary, we have scarcely explored even the outskirts of it. The "History of a Grain of Salt," "A Butterfly's Life," or "The Labours of an Ant," would introduce us to fairies and wonders quite as interesting as those of which we have spoken in these Lectures. While "A Flash of Lightning," "An Explosion in a Coal-mine," or "The Eruption of a Volcano," would bring us into the presence of terrible giants known and dreaded from time immemorial.

But at least we have passed through the gates, and have learnt that there is a world of wonder which we may visit if we will; and that it lies quite close to us, hidden in every dewdrop and gust of wind, in every brook and valley, in every little plant or animal. We have only to stretch out our hand and touch them with the wand of inquiry, and they will answer us and reveal the fairy forces which guide and govern them; and thus pleasant and happy thoughts may be conjured up at any time, wherever we find ourselves, by simply calling upon nature's fairies and asking them to speak to us. Is it not strange, then, that people should pass them by so often without a thought, and be content to grow up ignorant of all the wonderful powers ever active in the world around them?

Neither is it pleasure alone which we gain by a study of nature. We cannot examine even a tiny sunbeam, and picture the minute waves of which it is composed, travelling incessantly from the sun, without being filled with wonder and awe at the marvellous activity and power displayed in the infinitely small as well as in the infinitely great things of the universe. We cannot become familiar with the facts of gravitation, cohesion, or crystallization, without realizing that the laws of nature are fixed, orderly, and constant, and will repay us with failure or success according as we act ignorantly or wisely; and thus we shall begin to be afraid of leading careless, useless, and idle lives. We cannot watch the working of the fairy "life" in the primrose or the bee, without learning that living beings as well as inanimate things are governed by these same laws of nature; nor can we contemplate the mutual adaptation of bees and flowers without acknowledging that it teaches the truth that those succeed best in life who, whether consciously or unconsciously, do their best for others.

And so our wanderings in the Fairy-land of Science will not be wasted, for we shall learn how to guide our own lives, while we cannot fail to see that the forces of nature, whether they are apparently mechanical, as in gravitation or heat; or intelligent, as in living beings, are one and all the voice of the Great Creator, and speak to us of His Nature and His Will.

Freeriver
Community

Arabella Burton Buckley (1840 - 1929

Made in United States
Troutdale, OR
03/27/2025